Sentiment Analysis for Social Media

Sentiment Analysis for Social Media

Special Issue Editors

Carlos A. Iglesias
Antonio Moreno

MDPI • Basel • Beijing • Wuhan • Barcelona • Belgrade • Manchester • Tokyo • Cluj • Tianjin

Special Issue Editors

Carlos A. Iglesias
Departamento de Ingeniería de
Sistemas Telemáticos,
ETSI Telecomunicación
Spain

Antonio Moreno
Departament d'Enginyeria
Informàtica i Matemàtiques,
Escola Tècnica Superior
d'Enginyeria, Universitat Rovira
i Virgili (URV)
Spain

Editorial Office
MDPI
St. Alban-Anlage 66
4052 Basel, Switzerland

This is a reprint of articles from the Special Issue published online in the open access journal *Applied Sciences* (ISSN 2076-3417) (available at: https://www.mdpi.com/journal/applsci/special_issues/Sentiment_Social_Media).

For citation purposes, cite each article independently as indicated on the article page online and as indicated below:

LastName, A.A.; LastName, B.B.; LastName, C.C. Article Title. *Journal Name* **Year**, *Article Number*, Page Range.

ISBN 978-3-03928-572-3 (Pbk)
ISBN 978-3-03928-573-0 (PDF)

© 2020 by the authors. Articles in this book are Open Access and distributed under the Creative Commons Attribution (CC BY) license, which allows users to download, copy and build upon published articles, as long as the author and publisher are properly credited, which ensures maximum dissemination and a wider impact of our publications.

The book as a whole is distributed by MDPI under the terms and conditions of the Creative Commons license CC BY-NC-ND.

Contents

About the Special Issue Editors vii

Carlos A. Iglesias and Antonio Moreno
Sentiment Analysis for Social Media
Reprinted from: *Appl. Sci.* **2019**, *9*, 5037, doi:10.3390/app9235037 1

Hyoji Ha, Hyunwoo Han, Seongmin Mun, Sungyun Bae, Jihye Lee and Kyungwon Lee
An Improved Study of Multilevel Semantic Network Visualization for Analyzing Sentiment
Word of Movie Review Data
Reprinted from: *Appl. Sci.* **2019**, *9*, 2419, doi:10.3390/app9122419 5

Hannah Kim and Young-Seob Jeong
Sentiment Classification Using Convolutional Neural Networks
Reprinted from: *Appl. Sci.* **2019**, *9*, 2347, doi:10.3390/app9112347 31

Xingliang Mao, Shuai Chang, Jinjing Shi, Fangfang Li and Ronghua Shi
Sentiment-Aware Word Embedding for Emotion Classification
Reprinted from: *Appl. Sci.* **2019**, *9*, 1334, doi:10.3390/app9071334 45

Mohammed Jabreel and Antonio Moreno
A Deep Learning-Based Approach for Multi-Label Emotion Classification in Tweets
Reprinted from: *Appl. Sci.* **2019**, *9*, 1123, doi:10.3390/app9061123 59

Eline M. van den Broek-Altenburg and Adam J. Atherly
Using Social Media to Identify Consumers' Sentiments towards Attributes of Health Insurance
during Enrollment Season
Reprinted from: *Appl. Sci.* **2019**, *9*, 2035, doi:10.3390/app9102035 75

Sunghee Park and Jiyoung Woo
Gender Classification Using Sentiment Analysis and Deep Learning in a Health Web Forum
Reprinted from: *Appl. Sci.* **2019**, *9*, 1249, doi:10.3390/app9061249 85

Hui Liu, Yinghui Huang, Zichao Wang, Kai Liu, Xiangen Hu and Weijun Wang
Personality or Value: A Comparative Study of Psychographic Segmentation Based on an Online
Review Enhanced Recommender System
Reprinted from: *Appl. Sci.* **2019**, *9*, 1992, doi:10.3390/app9101992 97

Guadalupe Obdulia Gutiérrez-Esparza, Maite Vallejo-Allende and José Hernández-Torruco
Classification of Cyber-Aggression Cases Applying Machine Learning
Reprinted from: *Appl. Sci.* **2019**, *9*, 1828, doi:10.3390/app9091828 125

About the Special Issue Editors

Carlos A. Iglesias (Professor). Prof. Carlos Iglesias Associate Professor at the Universidad Politécnica de Madrid. He holds a Ph.D. in Telecommunication Engineering. He was previously Deputy Director at Grupo Gesfor and Innovation Director at Germinus XXI. He has been actively involved in research projects funded by private companies as well as national and European programs. His research interests are focused on intelligent systems (knowledge engineering, multi-agent systems, machine learning, and natural language processing).

Antonio Moreno (Professor). Dr. Moreno is a Full Professor of Artificial Intelligence at University Rovira i Virgili (URV) in Tarragona, Spain. He was the founder and director of the ITAKA (Intelligent Technologies for Advanced Knowledge Acquisition) research group until 2019. Since 2018, he has been the Deputy Director of URV Engineering School. He has been the author of more than 60 journal papers and over 125 conference papers. He has supervised 8 Ph.D. theses on different topics, including ontology learning, agents applied in health care, intelligent data analysis applied on healthcare data, recommender systems, and multi-criteria decision making. His current research interests are focused on sentiment analysis, recommender systems, and multi-criteria decision support systems.

Editorial

Sentiment Analysis for Social Media

Carlos A. Iglesias [1,*,†] and Antonio Moreno [2,†]

1. Intelligent Systems Group, ETSI Telecomunicación, Avda. Complutense 30, 28040 Madrid, Spain
2. Intelligent Technologies for Advance Knowledge Acquisition (ITAKA) Group, Escola Tècnica Superior d'Enginyeria, Departament d'Enginyeria Informàtica i Matemàtiques, Universitat Rovira i Virgili, 43007 Tarragona, Spain; antonio.moreno@urv.cat
* Correspondence: carlosangel.iglesias@upm.es; Tel.: +34-910671900
† These authors contributed equally to this work.

Received: 31 October 2019; Accepted: 19 November 2019; Published: 22 November 2019

Abstract: Sentiment analysis has become a key technology to gain insight from social networks. The field has reached a level of maturity that paves the way for its exploitation in many different fields such as marketing, health, banking or politics. The latest technological advancements, such as deep learning techniques, have solved some of the traditional challenges in the area caused by the scarcity of lexical resources. In this Special Issue, different approaches that advance this discipline are presented. The contributed articles belong to two broad groups: technological contributions and applications.

Keywords: sentiment analysis; emotion analysis; social media; affect computing

1. Introduction

Sentiment analysis technologies enable the automatic analysis of the information distributed through social media to identify the polarity of posted opinions [1]. These technologies have been extended in the last years to analyze other aspects, such as the stance of a user towards a topic [2] or the users' emotions [3], even combining text analytics with other inputs, including multimedia analysis [4] or social network analysis [5].

This Special Issue "Sentiment Analysis for Social Media" aims to reflect recent developments in sentiment analysis and to present new advances in sentiment analysis that enable the development of future sentiment analysis and social media monitoring methods. The following sections detail the selected works in the development of new techniques as well as new applications.

2. New Paths in Sentiment Analysis on Social Media

Traditionally, sentiment analysis has focused on text analysis using Natural Language Processing and feature-based Machine Learning techniques. The advances in disciplines such as Big Data and Deep Learning technologies have impacted and benefited the evolution of the field. This Special Issue includes four works that propose novel techniques.

In the first work, titled *"An Improved Study of Multilevel Semantic Network Visualization for Analyzing Sentiment Word of Movie Review Data"* [6], Ha et al. propose a method for sentiment visualization in massive social media. For this purpose, they design a multi-level sentiment network visualization mechanism based on emotional words in the movie review domain. They propose three visualization methods: a heatmap visualization of the semantic words of every node, a two-dimensional scaling map of semantic word data, and a constellation visualization using asterism images for each cluster of the network. The proposed visualizations have been used as a recommender system that suggest movies with similar emotions to the previously watched ones. This novel idea of recommending contents based on similar emotional patterns can be applied to other social networks.

In the second contribution, titled *"Sentiment Classification Using Convolutional Neural Networks"* [7], Kim and Jeong deal with the problem of textual sentiment classification. They propose a Convolutional Neural Network (CNN) model consisting of an embedding layer, two convolutional layers, a pooling layer, and a fully-connected layer. The model is evaluated in three datasets (movie review data, customer review data and Stanford Sentiment Treebank data) and compared with traditional Machine Learning models and state of the art Deep Learning models. Their main conclusion is that the use of consecutive convolutional layers is effective for relatively long texts.

In the third work, titled *"Sentiment-Aware Word Embedding for Emotion Classification"* [8], Mao et al. suggest the use of a sentiment-aware word embedding for improving emotional analysis. The proposed method builds a hybrid representation that combines emotional word embeddings based on an emotional lexicon with semantic word embeddings based on Word2Vec [9]. They use the emotional lexicon DUTIR, which is a Chinese ontology resource collated and labeled by the Dalian University of Technology Information Retrieval Laboratory [10]. This resource annotates lexicon entries with a model of seven emotions (happiness, trust, anger, sadness, fear, disgust and surprise). The evaluation is done with data from Weibo, a popular Chinese social networking site. The paper evaluates two methods (direct combination and addition) for building the hybrid representation in several datasets. They conclude that the experiments prove that the use of hybrid word vectors is effective for supervised emotion classification, improving significantly the classification accuracy.

Finally, in the fourth theoretical contribution, titled *"A Deep Learning-Based Approach for Multi-Label Emotion Classification in Tweets"* [11], Jabreel and Moreno address the problem of multi-class emotion classification based on Deep Learning techniques. The most popular approach for this problem is to transform it into multiple binary classification problems, one for each emotion class. This paper proposes a new transformation approach, so-called *xy-pair-set*, that transforms the original problem into just one binary classification problem. The transformation problem is solved with a Deep Learning-based system, so-called *BNet*. This system consists of three modules: an embedding module that uses three embedding models and an attention function, an encoding module based on Recurrent Neural Networks (RNNs), and a classification module that uses two feed-forward layers with the *ReLU* activation function followed by a sigmoid unit. The system is evaluated using the dataset *"Affect in Tweets"* of SemEval-2019 Task 1 [2], and it outperformed the state of the art systems.

3. Applications of Sentiment Analysis in Social Media

The wide range of applications of sentiment analysis has fostered its evolution. Sentiment analysis techniques have enabled to make sense of big social media data to make more informed decisions and understand social events, product marketings or political events. Four works selected in this Special Issue deal with the application of sentiment analysis for improving health insurances, understanding AIDS patients, e-commerce user profiling and cyberagression detection.

In the first work, titled *"Using Social Media to Identify Consumers' Sentiments towards Attributes of Health Insurance during Enrollment Season"* [12], van den Broek-Altenburg and Atherly aim at understanding the consumers' sentiments towards health insurances. For this purpose, they mined Twitter discussions and analyzed them using a dictionary-based approach using the NRC Emotion Lexicon [13], which provides for each word its polarity as well as its related emotion (anger, anticipation, disgust, fear, joy, sadness, surprise and trust). The main finding of this study is that consumers are worried about providers networks, prescription drug benefits and political preferences. In addition, consumers trust medical providers but fear unexpected events. These results suggest that more research is needed to understand the origin of the sentiments that drive consumers so that insurers can provide better insurance plans.

In the second contribution, titled *"Gender Classification Using Sentiment Analysis and Deep Learning in a Health Web Forum"* [14], Park and Woo deal also with the application of sentiment analysis techniques to health-related topics. In particular, they apply sentiment analysis for identifying gender in health forums based on Deep Learning techniques. The authors analyze messages from an AIDS-related bulletin board from HealthBoard.com and evaluate both traditional and Deep Learning techniques for gender classification.

In the third approach [15], titled *"Personality or Value: A Comparative Study of Psychographic Segmentation Based on an Online Review Enhanced Recommender System"*, Liu et al. analyze the predictive and explanatory capability of psychographic characteristics in e-commerce user preferences. For this purpose, they construct a pychographic lexicon based on seed words provided by psycholinguistics that are expanded using synonyms from WordNet [16], resulting in positive and negative lexicons for two psychographic models, Schwartz Value Survey (SVS) [17] and Big Five Factor (BFF) [18]. Then they construct word embeddings using Word2Vec [9] and extend the corpus with word embeddings from an Amazon corpus [19]. Finally, they incorporate the lexicons in a deep neural network-based recommender system to predict the users' online purchasing behaviour. They also evaluate customer segmentation based on BDSCAN clustering [20], but this does not provide a significant improvement. The main insight of this research is that psychographic variables improve the explanatory power of e-consumer preferences, but their prediction capability is not significant.

Finally, in the fourth work [21], titled *"Classification of Cyber-Aggression Cases Applying Machine Learning"*, Gutiérrez-Esparza et al. deal with the detection of cyberagression. They build and label a corpus of cyberagression news from Facebook in Latinamerica and develop a classification model based on Machine Learning techniques. The developed corpus can foster research in this field, given the scarcity of lexical resources in languages different from English.

4. Conclusions

The diversity of approaches of the articles included in this Special Issue shows the great interest and dynamism of this field. Moreover, this Special Issue of Applied Sciences contributes to provide a good overview of some of the main areas of research in this field.

Funding: This research received no external funding.

Acknowledgments: The Guest Editors would like to thank all the authors that have participated in this Special Issue and also the reference contact in MDPI, Nyssa Yuan, for all the support and work dedicated to the success of this Special Issue.

Conflicts of Interest: The authors declare no conflict of interest.

References

1. Liu, B. Sentiment analysis and opinion mining. *Synth. Lect. Hum. Lang. Technol.* **2012**, *5*, 1–167.
2. Mohammad, S.; Bravo-Marquez, F.; Salameh, M.; Kiritchenko, S. Semeval-2018 task 1: Affect in tweets. In Proceedings of the 12th International Workshop on Semantic Evaluation, New Orleans, LA, USA, 5–6 June 2018; pp. 1–17.
3. Cambria, E.; Poria, S.; Hussain, A.; Liu, B. Computational Intelligence for Affective Computing and Sentiment Analysis [Guest Editorial]. *IEEE Comput. Intell. Mag.* **2019**, *14*, 16–17.
4. Li, Z.; Fan, Y.; Jiang, B.; Lei, T.; Liu, W. A survey on sentiment analysis and opinion mining for social multimedia. *Multimed. Tools Appl.* **2019**, *78*, 6939–6967.
5. Sánchez-Rada, J.F.; Iglesias, C.A. Social context in sentiment analysis: Formal definition, overview of current trends and framework for comparison. *Inf. Fusion* **2019**, *52*, 344–356.
6. Ha, H.; Han, H.; Mun, S.; Bae, S.; Lee, J.; Lee, K. An Improved Study of Multilevel Semantic Network Visualization for Analyzing Sentiment Word of Movie Review Data. *Appl. Sci.* **2019**, *9*, 2419. [CrossRef]
7. Kim, H.; Jeong, Y.S. Sentiment Classification Using Convolutional Neural Networks. *Appl. Sci.* **2019**, *9*, 2347. [CrossRef]
8. Mao, X.; Chang, S.; Shi, J.; Li, F.; Shi, R. Sentiment-Aware Word Embedding for Emotion Classification. *Appl. Sci.* **2019**, *9*, 1334. [CrossRef]
9. Mikolov, T.; Chen, K.; Corrado, G.; Dean, J. Efficient estimation of word representations in vector space. *arXiv* **2013**, arXiv:1301.3781.

10. Chen, J. The Construction and Application of Chinese Emotion Word Ontology. Master's Thesis, Dailian University of Technology, Dalian, China, 2008.
11. Jabreel, M.; Moreno, A. A Deep Learning-Based Approach for Multi-Label Emotion Classification in Tweets. *Appl. Sci.* **2019**, *9*, 1123. [CrossRef]
12. Van den Broek-Altenburg, E.M.; Atherly, A.J. Using Social Media to Identify Consumers' Sentiments towards Attributes of Health Insurance during Enrollment Season. *Appl. Sci.* **2019**, *9*, 2035. [CrossRef]
13. Mohammad, S.M.; Kiritchenko, S.; Zhu, X. NRC-Canada: Building the state-of-the-art in sentiment analysis of tweets. *arXiv* **2013**, arXiv:1308.6242.
14. Park, S.; Woo, J. Gender Classification Using Sentiment Analysis and Deep Learning in a Health Web Forum. *Appl. Sci.* **2019**, *9*, 1249. [CrossRef]
15. Liu, H.; Huang, Y.; Wang, Z.; Liu, K.; Hu, X.; Wang, W. Personality or Value: A Comparative Study of Psychographic Segmentation Based on an Online Review Enhanced Recommender System. *Appl. Sci.* **2019**, *9*, 1992. [CrossRef]
16. Miller, G.A. WordNet: A lexical database for English. *Commun. ACM* **1995**, *38*, 39–41. [CrossRef]
17. Sagiv, L.; Schwartz, S.H. Cultural values in organisations: Insights for Europe. *Eur. J. Int. Manag.* **2007**, *1*, 176–190. [CrossRef]
18. McCrae, R.R.; Costa, P.T., Jr. The five-factor theory of personality. In *Handbook of Personality: Theory and Research*; The Guilford Press: New York, NY, USA, 2008; pp. 159–181.
19. McAuley, J.; Targett, C.; Shi, Q.; Van Den Hengel, A. Image-based recommendations on styles and substitutes. In Proceedings of the 38th International ACM SIGIR Conference on Research and Development in Information Retrieval, Santiago, Chile, 9–13 August 2015; pp. 43–52.
20. Ester, M.; Kriegel, H.P.; Sander, J.; Xu, X.. A density-based algorithm for discovering clusters in large spatial databases with noise. In *KDD-96 Proceddings*; AAAI Press: Portland, OR, USA, 1996; pp. 226–231.
21. Gutiérrez-Esparza, G.O.; Vallejo-Allende, M.; Hernández-Torruco, J. Classification of Cyber-Aggression Cases Applying Machine Learning. *Appl. Sci.* **2019**, *9*, 1828. [CrossRef]

© 2019 by the authors. Licensee MDPI, Basel, Switzerland. This article is an open access article distributed under the terms and conditions of the Creative Commons Attribution (CC BY) license (http://creativecommons.org/licenses/by/4.0/).

Article

An Improved Study of Multilevel Semantic Network Visualization for Analyzing Sentiment Word of Movie Review Data

Hyoji Ha [1], Hyunwoo Han [1], Seongmin Mun [1,2], Sungyun Bae [1], Jihye Lee [1] and Kyungwon Lee [3,*]

1. Lifemedia Interdisciplinary Program, Ajou University, Suwon 16499, Korea; hjha0508@ajou.ac.kr (H.H.); ainatsumi@ajou.ac.kr (H.H.); stat34@ajou.ac.kr (S.M.); roah@ajou.ac.kr (S.B.); alice0428@ajou.ac.kr (J.L.)
2. Science of Language, MoDyCo UMR 7114 CNRS, University Paris Nanterre, 92000 Nanterre, France
3. Department of Digital Media, Ajou University, Suwon 16499, Korea
* Correspondence: kwlee@ajou.ac.kr

Received: 30 April 2019; Accepted: 31 May 2019; Published: 13 June 2019

Abstract: This paper suggests a method for refining a massive amount of collective intelligence data and visualizing it with a multilevel sentiment network in order to understand the relevant information in an intuitive and semantic way. This semantic interpretation method minimizes network learning in the system as a fixed network topology only exists as a guideline to help users understand. Furthermore, it does not need to discover every single node to understand the characteristics of each clustering within the network. After extracting and analyzing the sentiment words from the movie review data, we designed a movie network based on the similarities between the words. The network formed in this way will appear as a multilevel sentiment network visualization after the following three steps: (1) design a heatmap visualization to effectively discover the main emotions on each movie review; (2) create a two-dimensional multidimensional scaling (MDS) map of semantic word data to facilitate semantic understanding of network and then fix the movie network topology on the map; (3) create an asterism graphic with emotions to allow users to easily interpret node groups with similar sentiment words. The research also presents a virtual scenario about how our network visualization can be used as a movie recommendation system. We next evaluated our progress to determine whether it would improve user cognition for multilevel analysis experience compared to the existing network system. Results showed that our method provided improved user experience in terms of cognition. Thus, it is appropriate as an alternative method for semantic understanding.

Keywords: collaborative schemes of sentiment analysis and sentiment systems; review data mining; semantic networks; sentiment word analysis

1. Introduction

1.1. Background and Purpose of Research

At present, we are faced with an enormous amount of information every day due to the vast growth of information and communications technology. Thus, there is increased interest in effective data processing and analysis. In particular, big data is playing an increasingly important role since it is suitable for refined and semantic processing, even if the amount of data is considerable or if its structure is complex [1]. Big data has also attracted great attention in the field of data visualization primarily for the design of efficient processing and semantic analysis. Data visualization is a redesigned concept of data analysis with better readability, offering distinct insights that cannot be grasped from a table or graph [2]. Network visualization is a visual tool to semantically analyze data if there is a massive amount or if the structure is complex [3]. Therefore, this study aims to demonstrate massive collective intelligence data through network visualization. This study also proposes an intuitive and semantic analysis method by designing multilevel sentiment network visualization based upon emotion words [4]. Social network analysis plays a significant role in understanding and finding solutions for society-functional problems by examining the original structure and relationships of the network. Therefore, network visualization is applied in a wide variety of fields, including network analysis based on data similarity, network analysis about social-scientific situations, graph theory, and recommendation systems.

Force-directed graph drawing algorithm is a standard layout algorithm for designing a network graph. It is considered to be highly useful since it allows related nodes to form a cluster [5]. However, the location of the node varies each time the graph is formed if a force-directed graph drawing algorithm is used because the entry value of the node's location is random and the eventual position is determined by relative connections between nodes. Therefore, users must repeat the learning of the system if in a force-directed network, since the absolute location information is not fixed (Figure 1). This is a notable drawback. Such difficulties can become great obstacles when interpreting the network if it consists of a considerable amount of data. Furthermore, the collective intelligence can deliver the wrong data if it is visualized based on a force-directed layout since locations of nodes may vary.

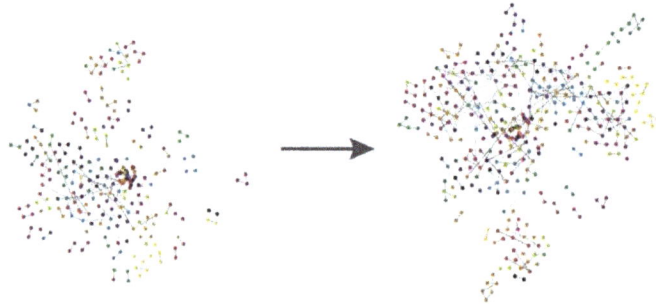

Figure 1. Force-directed layout network (**left:** 300 nodes; **right:** 400 nodes). Locations of nodes continue to change whenever data is added or modified. Reproduced with permission from [6]; published by [IEEE], 2015.

Following a preliminary study of ours [6–8] which introduced the sentiment network visualization approaches that are the basis of this work, we designed a multilevel sentiment network visualization so as to facilitate intuitive understanding of the complex collective intelligence data. Also, we present an approach to find solutions for those difficulties of a force-directed layout. Specifically, the primary contributions of our work are:

- More complete description of the sentiment movie network developing process: we present the explanation of flock algorithm method. Even if nodes increase in large scale, with this method, nodes won't be overlapped making bigger sentiment network configuration possible with improved delivery.
- Considering more diversity visual metaphorical graphic model, and conducting a survey about selection of graphic to better support interaction with the sentiment movie network.
- An evaluation for comparing the awareness level on network location.
- An evaluation for comparing between its final visualization map (Network + MDS+ Constellation Metaphor) and our previous visualization map [8].

1.2. Research Process

After targeting "movie review data" among examples of collective intelligence, we selected 36 sentiment words that generally appeared in movie reviews. These were classified into seven clustering characteristics through principal component analysis. These keywords were also analyzed through multidimensional scaling (MDS) to discover any correlations, such as similarities and differences. A two-dimensional distribution map was designed in the next step according to these correlations. We then designed a sentiment-based movie similarity network based on this two-dimensional distribution map through three steps. First, assuming that each node contains one piece of information about a movie, edges were created to match with each other if the nodes shared the most similar emotion characteristics. This eventually led to a network of 678 movie nodes. Heatmap visualization was also applied to allow us to more easily grasp the frequency of sentiment words of each movie [9]. A two-dimensional distribution map with sentiment words related to the movie review was created accordingly to distribute nodes in semantic positions. As each node was influenced by the semantic point depending on its attribute value, absolute positions of nodes were designed to reflect attributes of nodes [10]. Thus, nodes representing each movie formed a network layout which attracted nodes to where sentiment words were spread on the two-dimensional distribution map according to its frequency. The whole structure was referred to as the sentiment-movie network. Third, we applied a constellation visualization to label the characteristics of each cluster when nodes in a network structure showed clustering on a two-dimensional keyword distribution map. An asterism graphic consisted of objects representing traits of each clustering formed by nodes and edges of a network clustering structure. We next suggested a virtual scenario in the general public's view in order to determine how this network visualization could be applied in the actual movie recommendation process. Assuming that a random user enjoyed a certain movie, we demonstrated the process which the user went through in order to find a movie with similar emotions. Finally, three kinds of evaluations were conducted to verify whether the visualization method that we proposed could be linked to cognition improvement for users' multilevel analysis experience. The first test was designed to verify whether users would show a satisfactory understanding of the meaning of the location structure of nodes in a movie network visualization. The next test compared the two groups provided with or without heatmap visualization to see how well they could adopt sentiment words data when discovering the network. The last test was designed to determine which visualization case worked the most effectively for subject groups to conduct a semantic analysis among the following three cases: (1) the first visualization with network nodes; (2) the second visualization that involved fusing the first visualization with the two-dimensional sentiment word map indicating locations of the sentiments; and (3) the third visualization which superimposed the constellation metaphor over the second visualization. Figure 2 illustrates three processes for the sentiment analysis, data visualizing, and the evaluation workflow.

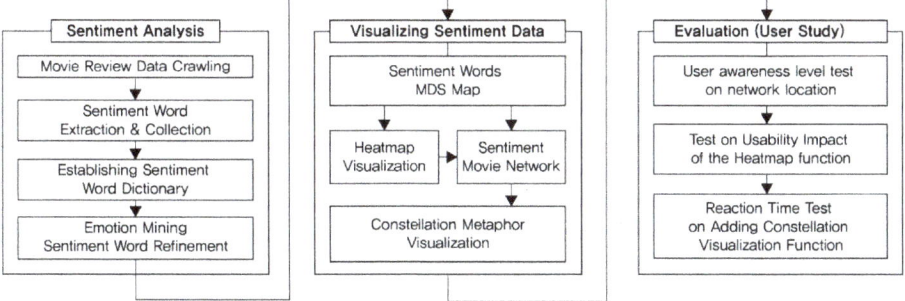

Figure 2. Research framework.

2. Materials and Methods

2.1. Sentiment Analysis and Visualization

Sentiment word analysis is a fundamental process in numerous applications of opinion and emotion mining, such as review crawling, data refining, and SNS classification [11]. In order to contribute to sentiment analysis and data visualization research fields, there is an abundance of studies that applied sentiment words and sentiment visualization system.

MyungKyu et al. [12] was used to deal with sentiment words shown in online postings and social media blogs. JoungYeon et al. [13] was used to illustrate adjectives used to describe the texture of haptic and indicated relations between adjectives on multi-dimensional scaling.

Ali et al. [14] provided a clear and logical taxonomy of sentiment analysis work. This taxonomy was made up of two main subcategories in this field: opinion mining and emotion mining. Also, they present a set of significant resources, including lexicons and datasets that researchers need for a polarity classification task. This would allow them to study the sentiment of any topic and determine the polarity of said topic, be it either positive or negative.

Kostiantyn et al. [15] suggest the state of the art sentiment visualization techniques and trends by providing the 132 cases of "sentimentvis" interactive survey browser. In this research, all the cases are classified based on the categorization standards. This categorization consists of the data domain, data source, data properties, analytic tasks, visualization tasks, visual variable, and visual representation. Also, the collected data indicates the growing multidisciplinary insight for visualization of sentiment with regard to multiple data domains and tasks.

2.2. Movie Recommendation

Studies on movie recommendation methods have mainly focused on "content-based recommendation systems utilizing information filtering technology" and "corporative recommendation system." According to Oard et al. [16], a content-based recommendation system was used to extract the characteristics from individual information and its following preference.

Movie recommendation systems based on corporative filtering have been analyzed by Sarwar et al. [17] and Li et al. [18]. They were used to recommend options selected by the group that shared similar information with individual users.

While these previous studies made recommendations based on individual information of users, we further managed user's experience data as in "emotional review data felt during the movie watching," thus enriching emotional attributes that fit the purpose of movie recommendation.

Recently, Ziani et al. [19] suggest a recommendation algorithm based on sentiment analysis to help users decide on products, restaurants, movies, and other services using online product reviews. The main goal of this study is to combine both the recommendation system and sentiment analysis in order to generate the most accurate recommendations for users.

However, this work has a limitation that did not present a user-centered recommendation system and focused on developing the automatic recommendation algorithm based on semi-supervised support vector machines (S3VM). Meanwhile, our system provides user-centered recommendation experience through using the network visualization and metaphorical sentiment graphic, which is easy to analyze.

2.3. Network Visualization and Layouts

A number of studies have been conducted on network visualization methods, including several recent studies on user's perception. For example, Cody Dunne et al. [20] have introduced a technique called motif simplification, in which common patterns of nodes and links are replaced with compact and meaningful glyphs, allowing users to analyze network visualization easily.

While this method identifies the maximal motif more accurately, even enabling the estimation of size through glyph and interaction, it has several difficulties for ordinary users. For example, users must put considerable effort toward learning concepts of motif and interpreting glyph. In addition, they have difficulty in discovering the optimal set of motifs.

Another study by Giorgio et al. [21] presented a tool called "Knot" with a focus on the analysis of multi-dimensional and heterogeneous data to achieve interface design and information visualization in a multidisciplinary research context.

Furthermore, Nathalie et al. [22] suggested methods to solve clustering ambiguity and increase readability in network visualization. That paper states that major challenges facing social network visualization and analysis include the lack of readability of resulting large graphs and the ambiguous assignment of actors shared among multiple communities to a single community.

They proposed using actor duplication in social networks in order to assign actors to multiple communities without substantially affecting the readability. Duplications significantly improve community-related tasks but can interfere with other graph readability tasks.

Their research provided meaningful insights as to how central actors could bridge the community. However, it also left confusions when distinguishing duplicated nodes and analyzing visualizations that exceeded a certain size since node duplications could artificially distort visualization.

Gloria et al. [23] present a novel method that uses semantic network analysis as an efficient way to analyze vaccine sentiment. This study enhanced understanding of the scope and variability of attitudes and beliefs toward vaccination by using the Gephi network layout.

These four studies mentioned above all aimed to improve network visualization from the perspective of users, with a particular focus on settling challenges of visualization distortion and existing network through users' learning of new technologies. However, network layouts used in previous studies could not handle overlapping of nodes when faced with increasing amounts of data.

In addition, the user may have to repeat the system since the location may vary. Furthermore, these network layouts were inefficient in that users had to check each node one by one in order to identify characteristics of a cluster.

Our research is consistent with previous studies in that it will fix the network data based upon sentiment words. It was designed to minimize users' learning and prevent distorted interpretation by applying a metaphor based on the characteristics of nodes in the network.

While previous studies made recommendations based on user's individual information, we further managed user's experience data as in "emotional review data felt during the movie watching," thus enriching emotional attributes that fit the purpose of movie recommendation.

3. Sentiment Analysis Data Processing

Sentiment analysis is the field of study that analyzes people's opinions, sentiment, attitudes, evaluations, survey, and emotions towards entities such as issues, events, topics, and their attributes. For sentiment analysis, we performed three data processes as follows.

3.1. Sentiment Words Collection

We selected 100 sentiment words filtered from 834 sentiment words based on the research conducted by DougWoong et al. [24] in order to create a sentiment word distribution map. A further survey of 30 subjects aged from 20 to 29 years determined the most frequently used words among these 100 sentiment words. Following basic instruction on the concept of sentiment words during movies, we investigated to what degree the emotion represented in each sentiment word could be drawn from watching some movies.

The survey began with the question "Describe how much you feel as in each sentiment words after watching the movies with the following genres, based on your previous experience." The questionnaire used a 7-point Likert Scale to evaluate responses ranging from "Strongly irrelevant" to "Strongly relevant." After eliminating 32 sentiment words that scored below the average, 68 sentiment words were finally selected [9].

3.2. Sentiment Words Refinement

In order to select the final sentiment words used for a two-dimensional distribution map among 68 sentiment words from the user survey, we collected and compared sentiment word data in existing movie reviews, eliminating words that were rarely used. This procedure consisted of three phases as shown below.

Crawling: Movie review data used in this research were collected from movie information service NAVER [25], a web portal site with the largest number of users in Korea. We designed a web crawler to automate sentiment word collection from movie reviews. This crawler covered three stages: (1) collecting unrefined movie reviews and tags in the NAVER movie web page, (2) refining collected data suitable for this research, and (3) extracting sentiment words based on analysis of refined data. As a result, we obtained 4,107,605 reviews of 2289 movies from 2004 to 2015.

Establishing sentiment word dictionary: We divided text data into morphemes collected through the crawling process using a mecab-ko-lucene-analyzer [26] and further extracted sentiment morphemes. A total of 133 morpheme clusters were selected through several text mining processes. A morpheme is the smallest grammatical unit in a language. In other words, it is the smallest meaningful unit of a language [27].

Each emotion morpheme selected was classified according to 68 kinds of detailed sentiment word categories. A sentiment word dictionary classified by the chosen sentiment word was then established. Extracting emotion morphemes and classifying them by category were conducted in consultation with Korean linguists.

To produce more accurate results, we eliminated less influential sentiment word clusters after matching them with actual movie review data. We calculated Term (w) Frequency (tf: Term Frequency) of each sentiment word cluster (t) suggested by the following formula.

$$tf(t,d) = \Sigma_{i=0}^{j} f(w_i, d) \qquad (1)$$

j = number of words in sentimental group t
The number of times that term t occurs in document d

Then, inverse document frequency (idf) was also drawn from this formula to lower the weight of the general sentiment word group. The inverse document frequency (idf) is "the logarithmically scaled inverse fraction of the documents that contain the word" [28]. It reduces the number of terms with a high incidence across documents.

$$idf(t, D) = \log\left(\frac{N}{|\{d \in D : t \in d\}|}\right) \qquad (2)$$

N = total number of documents in the corpus $N = |D|$
D = document set

TF-IDF score of sentiment word clusters on each movie was calculated using the following formula:

$$TFIDF(t, d, D) = tf(t, d) * idf(t, D) \qquad (3)$$

We next considered the maximum TF-IDF score that might appear from each sentiment word in order to decrease the number of sentiment words. For example, the word "Aghast" showed a TF-IDF score of no more than 0.8% in every movie, whereas "Sweet" scored 42% for at least one movie. We eliminated sentiment words with TF-IDF scores under 10%. Eventually, we selected 36 sentiment words. These sentiment word clusters were broadly divided into "Happy," "Surprise," "Boring," "Sad," "Anger," "Disgust," and "Fear" as shown in Table 1.

Table 1. Final sentiment words. Reproduced with permission from [6]; published by [IEEE], 2015.

Clustering Characteristics	Sentiment Words
Happy	Happy, Sweet, Funny, Excited, Pleasant, Fantastic, Gratified, Enjoyable, Energetic
Surprise	Surprised, Ecstatic, Awesome, Wonderful, Great, Touched, Impressed
Boring	Calm, Drowsy, Bored
Sad	Pitiful, Lonely, Mournful, Sad, Heartbroken, Unfortunate
Anger	Outraged, Furious
Disgust	Ominous, Cruel, Disgusted
Fear	Scared, Chilly, Horrified, Terrified, Creepy, Fearsome

3.3. Movie Data Collection

Movie samples used in network visualization were also collected from NAVER movie service in the same way as movie review data [25]. Based on 2289 movie samples from 2004 to 2015 registered in the NAVER movie service, movies with more than 1000 emotion morphemes were used to filter the emotion level. As a result, 678 movie samples were ultimately selected and used as network sample data.

4. Visualization Proposal

We proposed three methods to solve problems of the existing social network visualization layouts as follows.

4.1. Heatmap Visualization

Heatmap visualization is a method consisting of rectangular tiling shaded in scaled color [29,30]. It is used to search for anomalous patterns of data metrics as outlined in previous research [31]. To comprehend convergences, the method described by Young-Sik et al. [32] was used. A recent study proposed 100 cases of IT fractures, visualized and analyzed using the 2D parametric fracture probability heatmap [33]. The proposed map projection technique can project the high dimensional proximal femur fracture information onto a single 2D plane. Also, they applied heatmap in order to present the level of risk.

In this research, we showed TF-IDF size frequency of sentiment words in a heatmap, utilizing coordinate space in 2-dimensional distribution map of each sentiment word in order to visualize a sentimental distribution graphic for each movie node. The detailed research process is shown below. First, we measured space among the selected 36 sentiment words and analyzed correlations in order to design a two-dimensional sentiment word distribution map. We then conducted multi-dimensional scaling (MDS). We conducted a survey on the semantic distance among 36 sentiment words by enrolling 20 college students majoring in Digital Media Technology. These 36 sentiment words were placed on both axes (36 × 36) and the distance between words was scored in a range of plus/minus 3 points by considering their emotional distance. We used UNICET to facilitate a variety of network analysis methods based on data obtained from the 20 survey participants [10]. We also created Metric MDS as shown in Figure 3 based on the semantic distance among movie review semantic words. As a result, positive emotions such as "Happy" and "Surprise" were distributed on the right side of the X-axis while negative feelings such as "Anger" and "Disgust" were distributed on the left side. Active emotions that were generally exaggerated gestures such as "Fear" and "Surprise" were distributed on the top of the Y-axis while static emotions such as "Sad" and "Boring" were on the bottom. Furthermore, each type of sentiment words was clustered concretely based on particular characteristics such as "Happy," "Surprise," "Boring," "Sad," "Anger," "Disgust," and "Fear" on the two-dimensional distribution map. Results showed that cluster characteristic "Surprise" could be divided into "Happy" and "Fear" clusters. This implies that both overflowing joy and sudden fright are dominant in particular movies [10].

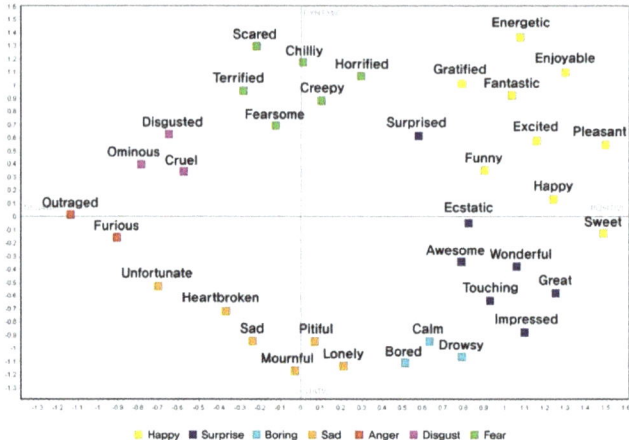

Figure 3. 36 sentiment words multidimensional scaling (MDS) map. Reproduced with permission from [6]; published by [IEEE], 2015.

Heatmap visualization was then designed based on a two-dimensional sentiment word distribution map as well as frequencies of these 36 sentiment words, which consisted of a two-dimensional distribution map. One optional movie was needed to design the heatmap. We measured frequencies of sentiment words on each movie by contrasting sentiment words in the movie review data obtained through the data construction process and sentiment words in the morphological dictionary. In addition, we measured numerical values by calculating the TF-IDF score to lower the weight of particular sentiment words that frequently emerged regardless of the typical characteristics of the movie. Therefore, TF-IDF score on each sentiment word could be interpreted as a numerical value reflected on the heat map visualization graph for target movies. The final heatmap graph consisted of a two-dimensional distribution map with sentiment words and tiny triangular cells. Every cell was initialized at numerical value, 0. The value then increased depending on the TF-IDF score of the sentiment word located in the pertinent cell. As the numerical value of the cell increased, the color of the cell changed, making it easier to discover the value of TF-IDF score of a pertinent sentiment word. Furthermore, as high-valued cells influenced the values of surrounding cells, the heatmap graph became gradually significant.

Figure 4b is a heatmap graph representing the distribution map of sentiment words from movie reviews written by viewers about the movie "Snowpiercer." This graph shows high frequencies of emotions such as "Pitiful and boring" as well as "Funny and great." Some reviews noted "Not so impressive, below expectation. It was pitiful and boring." "It was crueler than I thought, and it lasts pretty long in my head." and "The movie was pretty pitiful, probably because my expectations were too high." As shown in these reviews, there were various spectators with different emotions about this movie, including disappointment.

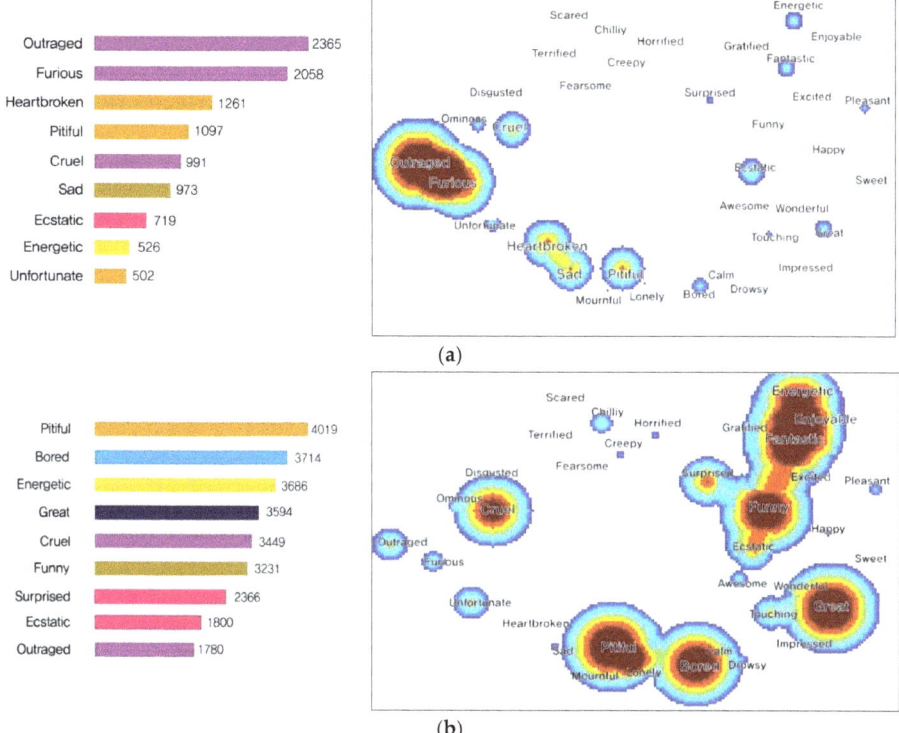

Figure 4. (a) Heatmap of "Don't Cry Mommy" showing single emotions (Furious, Outraged). (b) Heatmap of "Snowpiercer" that shows various emotions (Cruel, Pitiful, Lonely, Bored, Funny, Great, and Energetic). Reproduced with permission from [6]; published by [IEEE], 2015.

Therefore, a heatmap showing movie sentiment words can be divided into two cases. Thus, there are two types of heatmaps [7,8]. The first one is a type including movies with a single frequent sentiment word. The table shown above presented a ranking of sentiment words in a movie called "Don't cry mommy" (Figure 4a), while the table shown below is its actual movie reviews. Since the movie describes a sexual crime, emotions related to "Furious" and "Outraged" are prominent. Such parts are shown in red in the heatmap. The second is a type with various frequent sentiment words. For instance, the word frequency table and reviews on the movie "Snowpiercer" (Figure 4b) revealed different sentiment words such as "Boring, Cruel, Funny, and Pitiful." Thus, its heatmap suggests patterns based on such information. Furthermore, heatmap visualization helps us easily understand the types of emotions that viewers have experienced in a movie by reflecting the frequency of each sentiment word in the movie.

4.2. Sentiment Movie Network

In this session, we aim to describe the basic structures of suggested graphs and examples. Locations of nodes can be altered depending on the main sentiment word from the movie review. The suggested graph is similar to the artifact actor network, a type of multi-layered social network. the artifact actor network connects between the artifact network and social network using the semantic connection. Thus, it expresses the semantic relation between two networks [34]. In our proposed graph, we connected sentiment words on the 2-dimensional scaling map with movie network. In this paper, we referred to this network as the sentiment-movie network. Figure 5 shows the basic structure of the sentiment movie network [6].

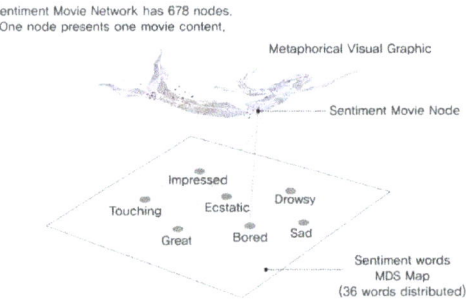

Figure 5. The multilevel structure of the sentiment movie network. Reproduced with permission from [6]; published by [IEEE], 2015.

As shown in Figure 5, the suggested graph is comprised of two layers. The first layer is called the semantic layer. It consists of semantic points based on the 36 sentiment words. The semantic point of the sentiment word is located at an initially set value. It is immovable. The second layer is called the network layer, which includes nodes comprising the movie network. Each movie node forms the edge of other movie nodes based on similarities. It also forms imaginary edges with the sentiment word in the two-dimensional distribution map based on the sentiment word connoted by the pertinent node. Nodes connected by edges have both attractive force and repulsive forces based on a force-directed algorithm. By contrast, semantic points of sentiment words are immovable, leaving only attractive force. For edge composition between nodes, we calculated cosine similarity between movies based on TF-IDF scores of the 36 sentiment words. The similarity between movie A and movie B or SIM (A, B) is shown as follows.

$$SIM(A, B) = \frac{\sum_{i=0}^{n} A_i * B_i}{\sqrt{\sum_{i=0}^{n} (A_i)^2} * \sqrt{\sum_{i=0}^{n} (B_i)^2}} \quad (4)$$

The edge between each node and the semantic point sets up a fixed threshold value and generates an edge by designating a sentiment word with a value that is greater than a threshold value as semantic feature. Figure 6a,b show an example that the location of a node on the graph can be altered depending on the frequency of sentiment word indicated in the Heatmap Visualization [7,8].

Figure 6a shows that the node is located in the space of the sentiment word with overwhelmingly high frequency. If a movie has more than two frequent sentiment words, its nodes will be located in the way, as shown in Figure 6b. If shown in the heatmap visualization of Figure 6b, nodes in "Snowpiercer" are influenced by keywords such as Pitiful, Bored, Energetic, Great, and Cruel. In addition, nodes are particularly influenced by locations of "pitiful" and "bored" since those two are the strongest that they affect nodes to be placed around the negative y-axis. As such, frequent sentiment words will place nodes.

To show users a host of nodes in a meaningful position without overlapping, flock algorithm has been applied. Flock algorithm consists of separation, alignment, and cohesion [35]. We composed nodes applying separation method because nodes of developed visualization do not move and have no direction.

Solution: One node compares all nodes to each other's positions and moves in opposite directions if overlaid.

	Algorithm 1. Separate nodes–avoid crowding neighbors
	Input: *nodes*→Node objects to be displayed on the screen.
	Output: Node objects whose locations have been recalculated so that they are distributed without overlap.
	Method
1:	Initialize SEPARATION_DISTANCE
2:	WHILE true
3:	change = false
4:	FOR *src* in *nodes*
5:	FOR *dst* in *nodes*
6:	IF *src* == *dst*
7:	continue
8:	ELSE
9:	dist = src.position − dst.poisiton
10:	IF dist.size < SEPARATION_DISTANCE
11:	change = true
12:	src.position += 1/dist * dist.direction
13:	dst.position −= 1/dist * dist.direction
14:	END IF
15:	END IF
16:	END FOR
17:	END FOR
18:	IF *change* == false
19:	break
20:	END WHILE

Even if nodes increase in large scale or similar nodes are added, with this method, nodes will not be overlapped making bigger network configuration possible with improved delivery.

As every node connected by the network made of suggested methods is located in the graph, clustering is formed by combining similar movies in the space of sentiment word with high frequency considering connections between movies as well as connections between related sentiment word. Figure 7 shows an extreme position of a node and a cluster [8]. This network allows users to easily understand the network structure even if the number of nodes is changed by fixating the topology of movie networks based on the sentiment word distribution map of movie reviews. Finally, k-means clustering operation using cosine similarity value for classifying cluster characteristics of each node was conducted. The number of clusters considered ranged from 9 to 12. The final cluster number was chosen to be 11 as the node number of each cluster was evenly distributed and various characteristics were well clustered. Furthermore, each node was colored to classify each node group based on these 11 clusters.

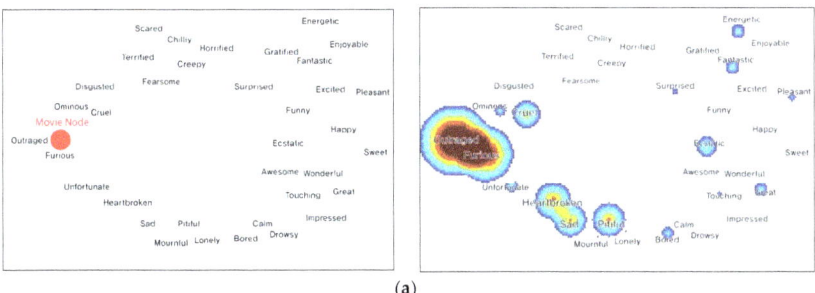

(a)

Figure 6. Cont.

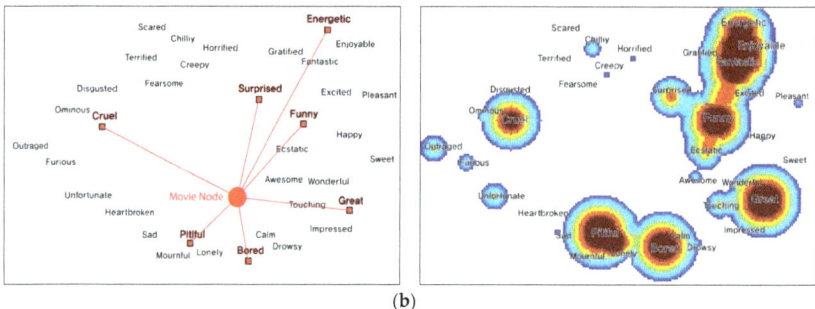

Figure 6. (**a**) Heatmap visualization and positioning on the sentiment-movie network. (one point position) in the case of "Don't Cry Mommy." (**b**) Heatmap visualization and positioning on the sentiment-movie network. (more than two point positions) in the case of "Snowpiercer." Reproduced with permission from [8]; published by [Journal of The Korea Society of Computer and Information], 2016.

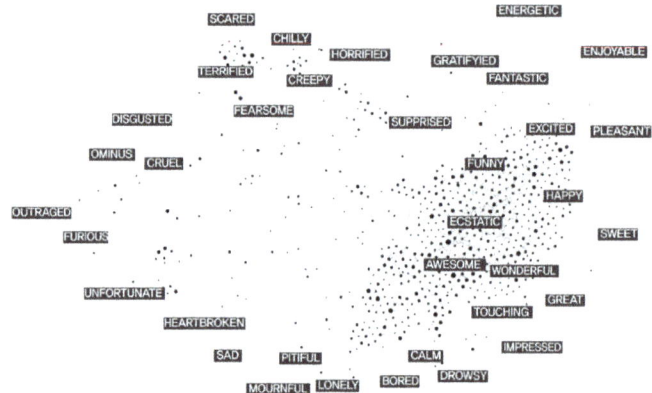

Figure 7. Sentiment movie network (678 Movie nodes). Reproduced with permission from [6]; published by [IEEE], 2015.

4.3. Constellation Visualization

This session facilitates a cognitive understanding of the process used to design constellation image visualization based on specific nodes and edges with significant sentiment word frequency in order to clarify semantic parts of each clustering. Metaphorical visualizations can map characteristics of certain well-understood visual images/patterns to a more poorly understood data source so that aspects of the target can be more understandable.

Recently, Information graphics and the field of information visualization apply a variety of metaphorical devices to make large size, complex, abstract, or otherwise difficult-to-comprehend information understandable in graphical terms [36,37].

We created an asterism graphic of each cluster network by considering the significant sentiment words, information on movies, and synopses in each cluster. Our research makes it easy to understand the sentiment data system to the general public by using the visual metaphor graphics.

In order to realize asterism images, we referred to the labeling data of the 11 different clusters yielded from k-means clustering, the most dominant categories of sentiment words in each cluster, and their following information on movies and synopsis. In order to select a pool of constellation graphics, experts with academic backgrounds in graphic design and movie scenario professionals were consulted, ultimately narrowing down options to 2~3 graphics per sentiment cluster. Table 2 below contains the lists of image options corresponding to each of these 11 clusters.

For the next phase, 30 students majoring in design and visual media were surveyed to choose the final constellation graphics for representing sentiment words around each cluster. The survey incorporated statements that described relations between sentiment words shown in clusters as well as the choice of images in a 5-point Likert scale. For instance, when a statement of "An image of a dynamite bears relation with the feeling of surprise" is presented, the test subject was supposed to mark an answer from a range of "not at all (1 point) ~ very much (5 points)" and evaluate the strength of the link between the sentiment word and the graphic. Table 2 presents the average drawn from these 30 students' data. The graphics with the highest average was chosen as the final constellation to represent the relevant cluster.

Table 2. List of candidate graphics linked to a cluster of sentiment words.

Cluster Name	List of Candidate Graphics
Cruel and dreadful	Red pyramid (4.7), Piranha (4.1), Jigsaw (3.3)
Dramatic Emotional	Comet (2.1), Ballerinas (3.5), **Whale (4.1)**
Dynamic mood change	Chameleon (3.3), **Persona mask (4.3)**
Thrilling and horrifying	**Alien (3.8)**, Jaws (3.0)
Surprising	A surprised face mask (1.7), Dynamite (3.4), **Jack in the box (4.2)**
Pleasing and exciting	Firecracker (3.3), **Gambit (3.7)**, Party hat (3.2)
Authentic fear	**Reaper (4.5)**, Scream Mask (4.1), Dracula (2.3)
Generally Monotonous	**Sloth (3.0)**, Snail (2.1), Yawner (2.3)
Fun and cheerful	**Wine Glass (4.3)**, Heart (3.2), A diamond ring (3.5)
Cute	**Gingerbread Cookie (3.5)**, Kitty (3.0)
Sad and touching	**Mermaid (4.2)**, Teardrop (2.8)

Table 3 below shows the main emotions and movie examples contained in each cluster as well as motives for choosing each asterism name [6]. This constellation visualization helped us naturally perceive characteristics of a node without having to individually review parts of the network.

Table 3. Definition of constellation visualization. Reproduced with permission from [6]; published by [IEEE], 2015.

Cluster Name	Movie Examples	Asterism Name	Motives for Each Name
Cruel and dreadful	Final Destination 3, Piranha 3D	Red pyramid	Symbolized the cruelly murdering character in a movie <Silent Hill>
Dramatic Emotional	Pride & Prejudice, The Notebook	Whale	Inspired from the scene when grampus appears in a movie <Life of Pi>, which aroused dramatic and emotional image simultaneously
Dynamic mood change	Snowpiercer, Transformers	Persona mask	Persona masks are supposed to express various emotions, which is similar to movies with dynamic mood changes
Thrilling and horrifying	Resident Evil, War Of The Worlds	Alien	Aliens arouse fear and suspense in unrealistic situations
Surprising	Saw, A Perfect Getaway	Jack in the Box	Symbolized an object popping out of the box to express surprising moments
Pleasing and exciting	Iron Man, Avatar	Gambit	Relevant to the magician character of a movie <X-men>, who is fun and surprising
Authentic fear	Paranormal Activity, The Conjuring	Reaper	Symbolized as a reaper to show the authentic and intrinsic fear
Generally Monotonous	127 Hours, Changeling	Sloth	Originated from the idea that sloths are boring and mundane
Fun and cheerful	Hairspray, The Spy: Undercover Operation	Wine Glass	Wine glass is a symbol of romantic and festive atmosphere
Cute	Despicable Me, Puss In Boots	Gingerbread Cookie	Gingerbread men cookies represent cute and sweet sensations
Sad and touching	Million Dollar Baby, Man on fire	Mermaid	The story of little mermaid shows touching, magical and sad ambience at the same time

Constellation visualization helped us naturally perceive characteristics of a node without having to review parts of the network individually. A comprehensive network map based on information presented in this table is shown in Figure 8. We call this movie data network based on three proposals "CosMovis." Figure 8 also indicates that it is substantially easier to semantically analyze network visualization using overlapping asterism images on each sentiment word and symbolic nodes with connection structure of edges.

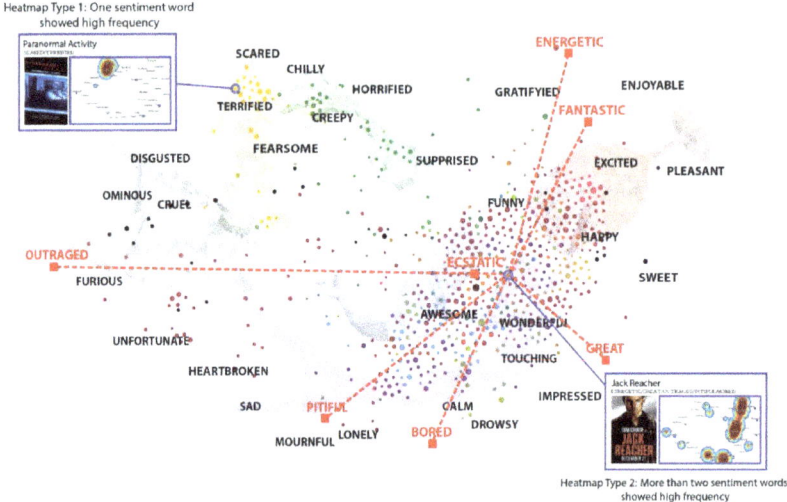

Figure 8. Comprehensive "CosMovis" constellation map of sentiment word-based movies. (Demo Site URL: https://hyunwoo.io/cosmovis/). Reproduced with permission from [7]; published by [IEEE], 2014.

5. CosMovis Network Analysis Scenario

Our research fixed the network data based upon sentiment words MDS map and was designed to minimize users' learning and prevent distorted interpretation, by applying a metaphor based on the sentiment keyword characteristics of the nodes in the network. In order to show that the user's target can be applied widely, this section presents an ideal scenario about how our sentiment network visualization is used as a movie recommendation and how it affects a general public's decision-making process [8]. The scenario is shown as follows.

Scenario: Getting a movie recommendation from the cluster with similar emotions based on previous movies that the general public has watched previously.

This scenario describes a situation of obtaining a movie recommendation from clusters that share similar emotions based on sentiment words implied by users' previous movie preferences. Assume that the user watched "Star Trek" and was satisfied with the emotions provided by the movie. The user will then search for a new movie from the network. He/she will first find nodes of "Star Trek" in the network and then look for movies closely located to "Star Trek." As a result, the user will be able to find three other movie nodes based on "Star Trek" as shown in red in Figure 9.

Next, the user will focus on the heatmap to understand what kinds of emotions those three movies contain. Figure 10 on the right shows heatmaps of movie nodes A, B, and C as well as their sentiment word distribution table. Node A is for the movie "Avatar" and Node B is for the movie "Ironman 3." One can see that heatmaps of these two nodes are very similar to "Star Trek." Thus, if the user wants to have a similar emotional experience, one could watch Avatar or Ironman 3. Node C is for the movie "Masquerade." According to its heatmap, this movie also contains "touching" aspects in addition to other emotions present in "Star Trek." Therefore, if one wants a movie more emotional than "Star Trek," the user could pick this movie.

We can also note that node C is for a historical genre movie, "Masquerade," unlike "Star Trek," "Avatar," or "Ironman 3". This implies that sentiment-based movie similarity network can also recommend movies in different genres as long as they contain similar sentiments. Likewise, users can make more effective decisions if they want to obtain movie recommendations based on movies they have watched by understanding the network structure, selecting candidate movies among ones similar to his/her previously enjoyed movies, and analyzing sentiment word frequency of each candidate node.

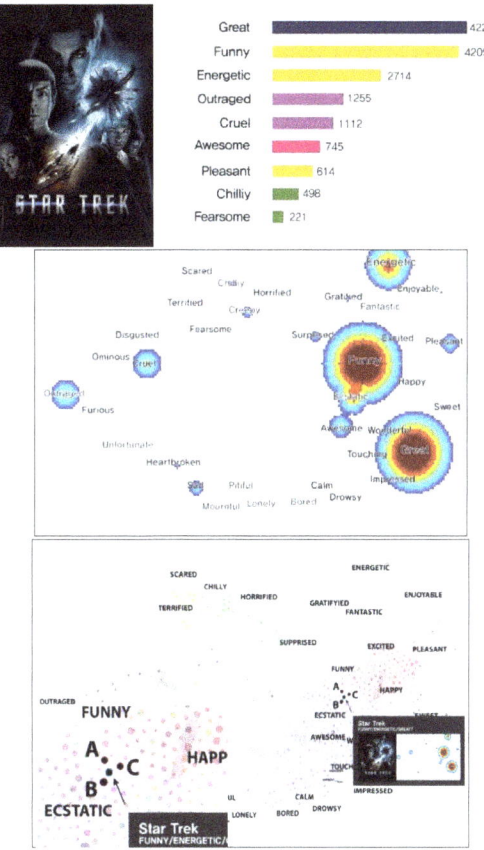

Figure 9. (**upper**) Movie "Star Trek" Poster and TF-IDF size frequency of sentiment words. (**middle**) "Star Trek" heatmap visualization. (**lower**) After selecting the movie "Star Trek (**blue node**)," three movies A, B, and C (**red nodes**) that are located closely within the same cluster are discovered. Reproduced with permission from [8]; published by [Journal of The Korea Society of Computer and Information], 2016.

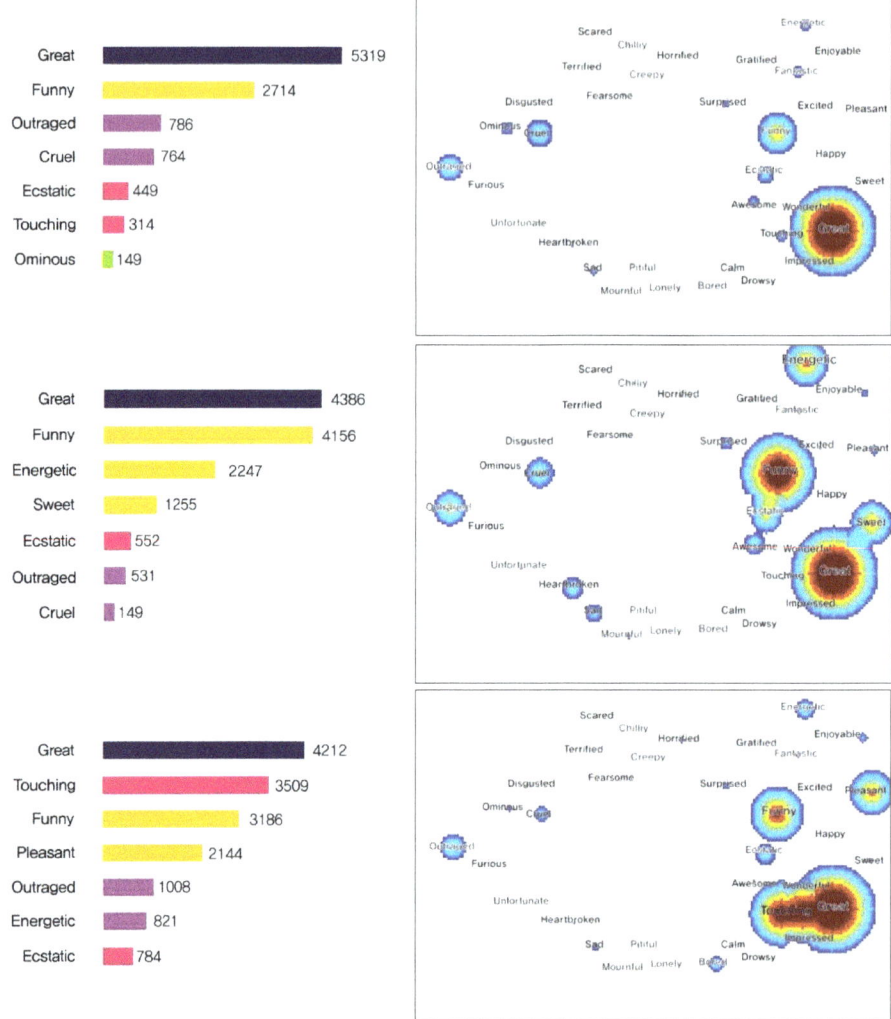

Figure 10. (**upper**) Movie "Avatar" TF-IDF size frequency of sentiment words and heatmap visualization. (**middle**) Movie "Ironman3," (**lower**) movie "Masquerade."

6. Evaluation

The scenario is as follows. Using an experimental method, we will explore whether the network visualization put forth by this research can improve user cognition. For the purpose of verification, three evaluations were conducted as follows. All evaluations were designed based on social science and statistics in accordance with the International Review Board regulations.

6.1. Evaluation 1: User Awareness Level Test on Network Location

Purpose and Method of Cognition Evaluation: To gauge the users' awareness level of the network structure created through this research, an evaluation involving 40 participants was designed. The participants were divided into two groups of 20, with one group given a simple explanation regarding the visualization for two minutes while the other group was provided with a deeper explanation for five minutes. After a single movie node was selected from the network and the corresponding heatmap shown, participants were asked to perform a series of tasks such as "Choose the heatmap located adjacent to (or across from) the selected node." The movie network was verified through three axes, as illustrated below (Figure 11). The survey contained a total of 20 questions, including five questions per axis along with an additional five questions pertaining to the node in the center of the network.

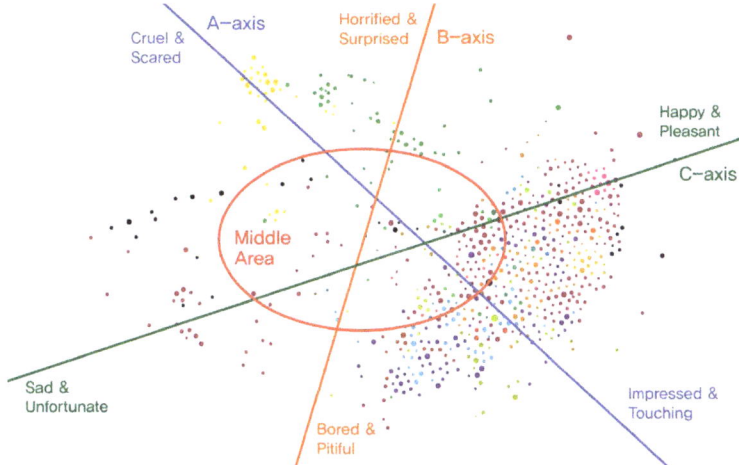

Figure 11. Three axis and sentiment word directional guide for verifying the network structure.

Comparison of group average: For the purpose of average comparison, an independent *t*-test was conducted. An independent *t*-test was also used to verify whether a difference in population means existed between two groups that followed normal distribution N (μ_1, σ_1^2) and N (μ_2, σ_2^2) The two groups were assumed to be independent of each other [38]. For this research, in order to discover the difference in awareness levels between the two groups given either a simple or a detailed explanation on the network structure, a survey asked participants to find the heatmap that represented a particular node on the visualization map. We recruited 40 participants (20 in each group) who were majoring in related fields with an understanding of visualization. We also assumed a normal distribution in accordance with the central limit theorem. Three results on these two groups regarding network structure are presented in Table 4.

Table 4. Results of independent *t*-test analysis.

Question	P-Value	Equal Variance Assumption	T-Value	P-Value	P-Value/Alternative Hypothesis Adoption
b_A~a_A	0.01079 **	Heteroscedasticity **	−2.1963	0.03612 **	Adopt **
b_B~a_B	0.8118	Equal variance	−2.5591	0.01461 **	Adopt **
b_C~a_C	0.3576	Equal variance	−1.9321	0.06117 *	Adopt *
b_M~b_M	0.6122	Equal variance	−1.3635	0.1809	Dismissal

* 90% Confidence level, ** 95% Confidence level (A: A-axis/B: B-axis/C: C-axis/M: Middle area)/ (a: 2 min—Group/ b: 5 min—Group)/ (Left~Right: Compare Left and Right).

Independent *t*-test results between the more-informed group and the less-informed group showed that *P*-values for questions in the three axes (excluding the middle area) were smaller than the alpha-value (0.05), which was the critical value of a 95% confidence level, or (0.1), which was the critical value of a 90% confidence level. This indicated a significant difference between the two groups that were more-instructed and less-instructed about the network. Table 5 and Figure 12 present details regarding the two groups compared through independent *t*-tests.

Table 5. Details regarding the two test groups.

Question	95% Confidence	Before	After
b_A~a_A	−1.351534< μ <−0.048465	3.9	4.6
b_B~a_B	−1.612026< μ <−0.187974	3.4	4.3
b_C~a_C	−1.741933< μ <−0.041933	2.5	3.35
b_M~a_M	−1.242694< μ <0.242694	2.15	2.65

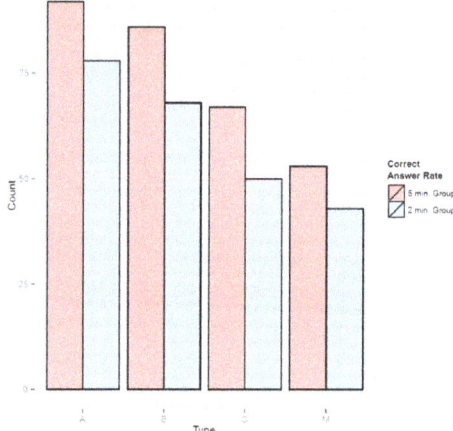

Figure 12. Comparison of correct answer rate between test groups.

Details on the two test groups indicated that, in terms of the rate of providing correct answers for questions on all axes, the well-informed group outperformed the less-informed group in terms of network structure. For the A-axis, after being provided a further explanation about the network structure, the percentage of getting correct answers increased by 0.7. Similarly, for B-Axis, C-Axis, and center node, percentages of getting correct answers increased by 0.9, 0.85, and 0.5, respectively. Accuracy improvement was most pronounced for the A-Axis, whereas questions regarding the center node produced the lowest enhancement in performance. Thus, we can infer that explaining the network structure in advance of relevant visualization use can help raise the level of user awareness on visualization. To further boost awareness regarding the center node, along with an explanation on the network structure, a supplementary method may be required. Nevertheless, with the correct answer rate for the less-informed group hovering around 50% on average, we can conclude that user awareness of visualization is considered high in general.

6.2. Evaluation 2: Test on Usability Impact of the Heatmap Function

Purpose and Method of Usability Measurement Experiment: The first part of this research concluded that explaining the network structure prior to visualization usage could enhance user awareness [8]. To further improve the awareness level on center nodes, an additional prop up measure was required. As one possible method for boosting user awareness on center nodes, this study introduced a new function, through which dragging the mouse over to a particular center node would offer heatmap visualization. The second test was thus designed to confirm whether a difference in visualization usability existed between group 1 (provided with the heatmap function) and group 2 (not provided such a function). The usability was measured through a survey comprised of five questionnaires regarding learnability, efficiency, understandability, feature functionality, and accuracy measured with a Likert scale of 7. Forty college students (two groups of 20) who were currently studying data visualization with knowledge of the visualization field were selected as samples.

Reliability Analysis: Before analyzing the data gathered, a reliability analysis of the above survey was carried out in order to verify its reliability. Reliability analysis measures internal consistency among questions/statements employed in the survey based on Cronbach's α. Cronbach's α has a value of between 0 and 1. The closer the value is to 1, the higher the question's reliability. Generally, a Cronbach's α of 0.6 or higher is considered to be reliable. This single index can be used for all questions to yield a comprehensive analysis. Results of reliability analysis performed on the final data set are presented in Table 6.

Table 6. Results of reliability analysis. Reproduced with permission from [8]; published by [Journal of The Korea Society of Computer and Information], 2016.

Categories	Statements	Cronbach's α (Provide)	Cronbach's α (Non-Provide)	Total Cronbach's α
Learnability	1. It is easy to select a movie based on the sentiment words.	0.698	0.827	
Efficiency	2. It is efficient to select the node based on the sentiment of the movie.	0.698	0.826	0.666
Understandability	3. It is easy to understand the sentiment distribution depending on varying node locations.	0.661	0.747	
Feature Functionality	4. It provides an adequate function to help user choose a movie based on certain sentiment words.	0.663	0.749	
Accuracy	5. The selected movie and the sentiment distribution predicted from the movie's map coordinate matches.	0.742	0.725	

Comparing maximal values of Cronbach's α after eliminating a certain item indicated that the number was the highest (0.742) after eliminating Accuracy if the heatmap was provided while it had the maximal value of 0.827 after eliminating Learnability if the heatmap was not provided. In addition, the Cronbach's α value exceeded 0.6, suggesting the absence of unreliability per each statement as well as a high level of internal consistency across the survey. Thus, it can be concluded that the survey is highly reliable.

Average Comparison per Group: In order to compare averages, an independent t-test was conducted. An independent t-test was also used to verify whether there was a difference in population means between two groups that followed a normal distribution $N(\mu_1, \sigma_1^2)$ and $N(\mu_2, \sigma_2^2)$. The two groups are assumed to be independent of each other [39]. This research, through further data distillation, examined 30 sets of survey data in accordance with the central limit theorem. The two groups were assumed to follow a normal distribution and be independent of each other in comparing averages. Results of average comparison analysis of the two groups are shown in Table 7.

Table 7. Results of the independent *t*-test analysis. Reproduced with permission from [8]; published by [Journal of The Korea Society of Computer and Information], 2016.

Question	P-Value	Equal Variance Assumption	T-Value	P-Value	P-Value/Alternative Hypothesis Adoption
1_1~1_2	0.08203	Heteroscedasticity	4.8295	0.00003 **	Adopt **
2_1~2_2	0.5064	Heteroscedasticity	7.2038	0.00000001 **	Adopt **
3_1~3_2	0.2327	Heteroscedasticity	4.7609	0.000032 **	Adopt **
4_1~4_2	0.07771	Heteroscedasticity	4.3814	0.00011 **	Adopt **
5_1~5_2	0.0026	Equal variance	4.9205	0.000036 **	Adopt *

* 90% Confidence level, ** 95% Confidence level (a: Heatmap group/b: No heatmap group)/ (Left~Right: Compare Left and Right).

After conducting an independent *t*-test on the two groups to which heatmap function was either available or not available, it revealed significant differences between the two. Table 8 below presents details of statements along with the alternative hypothesis.

Table 8. Details of statements with alternative hypothesis. Reproduced with permission from [8]; published by [Journal of The Korea Society of Computer and Information], 2016.

Question	95% Confidence	Provide	Non-Provide
1_1~1_2	1.128555< μ <2.771445	5.45	3.5
2_1~2_2	1.86879< μ <3.33121	5.95	3.35
3_1~3_2	1.118809< μ <2.781191	5.8	3.85
4_1~4_2	0.9908482< μ <2.7091518	5.65	3.8
5_1~5_2	1.254113< μ <3.045887	5.75	3.6

(Left~Right: Compare Left and Right).

For all statements that rejected the null hypothesis (providing heatmap function does not produce a significant difference between the two groups) and adopted the alternative hypothesis (providing heatmap function produces a significant difference between the two groups), it was confirmed that the average value of the group provided with the heatmap function was higher than that of the control group. Thus, we could conclude that a valid difference existed between the two groups based on whether or not heatmap function was provided. We could further conclude that providing the heatmap function reinforced the usability of the visualization map. It was also effective in enhancing the awareness level of middle area nodes, which required an additional measure to be understood better based on the first part of this research.

6.3. Evaluation 3: Reaction Time Test on Adding Constellation Visualization Function

Purpose and Method of Measuring Reaction Time: This research proceeds with a comparison between its final visualization map and previous visualization maps [6–8]. The final visualization in this study was formed by combining clustered (colored) nodes based on sentiment words, the sentiment word MDS map indicating locations of sentiments, and the constellation metaphor applied to sentiment words. This comparison test was designed to determine whether there was any difference in reaction time spent when selecting the node for each visualization. Visualizations applied in this test consisted of three different types: visualizations 1 to 3. Table 9, Table 10 and Figure 13 present further details of these types. Test subjects were divided into three groups, one for each version of the visualization. In this study, visualization scenes were divided into four quadrants. Subjects were given 30 s to observe each quadrant of the visualization. Thus, the total observation time was 120 s. Meanwhile, subjects were allowed to discover the content and structure of "CosMovis" freely. They were asked questions (e.g., on which part of the map is the movie cluster that stirs touching and sad feelings located?) on complex sentiments included in six polygons. Reaction time for recognizing the location of the complex sentiment was then measured. College students currently studying data visualization with knowledge of the visualization field were selected as samples. A total of 45 participants were tested (15 participants for each visualization group) [40].

Table 9. The basic structure of visualization (1 to 3).

Visualization Type	Basic Structure
Visualization 1 (Network)	Visualization on clustered (colored) nodes from sentiment words
Visualization 2 (Network + MDS)	Combined visualization on clustered nodes and MDS map which shows the location of sentiments
Visualization 3 (Network + MDS + Constellation Metaphor)	Combined visualization on clustered nodes, MDS map which shows the location of sentiments, and constellation metaphors based on sentiment words

Table 10. Polygon-related questions.

Polygon	Polygon Name (Sentiment Words)	Equal Variance Assumption
	Mermaid (Sad, Touching)	1. Where are the movies that stir up touching and sad feelings clustered around?
	Wine Glass (Funny, Sweet, Cheerful)	2. Where are the movies that arouse fun, sweet and pleasing emotions clustered around?
	Jack in the box (Surprising)	3. Where are the movies that trigger feelings of surprise clustered around?
	Red-Pyramid (Cruel, Dreadful)	4. Where are the movies that induce cruel and dreadful emotions clustered around?
	Reaper (Authentic fear)	5. Where are the movies that set off genuine fear clustered around?
	Whale (Dramatic Emotional)	6. Where are the movies that elicit dramatic sentiments clustered around?

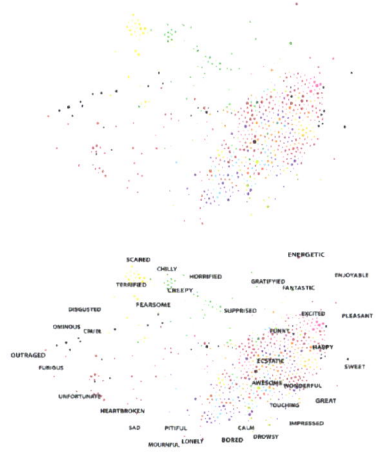

Figure 13. Cont.

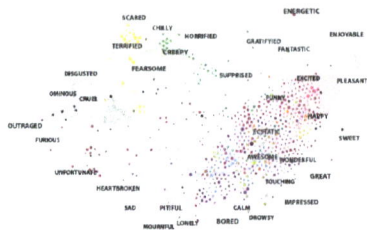

Figure 13. (**upper**) First visualization: network; (**middle**) second visualization: network + MDS (**lower**) third visualization: network + MDS + constellation metaphor.

ANOVA per group: To analyze the variance of data pertaining to multiple groups, this study conducted two-way ANOVA. Two-way ANOVA can be used when a data set is comprised of two independent variables and one dependent variable in order to verify the existence of a significant difference in variance among different groups caused by changes in an independent variable [41]. This study collected data through a closed testing method in which a series of 1:1 individual tests were undertaken. We used questionnaires and measured reaction times of the three groups that were given different versions of visualization for accurately locating the complex sentiment suggested by the six polygons. Results of two-way ANOVA are presented in Table 11.

Table 11. Results of the two-way ANOVA.

	DF	Sum Sq	Mean Sq	F-Value	Pr(>F)
Group	1	618.5	618.5	278.387	2×10^{-16} **
Question	1	83.8	83.8	37.74	2.16×10^{-9} **
Question: Group	1	0.1	0.1	0.056	0.814
Residuals	356	790.9	2.2		

* 90% Confidence level, ** 95% Confidence level.

We reviewed the difference in reaction time among groups given different versions of visualization maps and six kinds of polygon questions. The visualization map-based test result revealed an F-Value of 278.387 and a P-Value of 2×10^{-16} lower than alpha value of 0.05, leading us to discard the null hypothesis (no difference exists in reaction time to different polygon questions among groups) and adopt the alternative hypothesis (a difference exists in reaction time to different polygon questions among groups). In the polygon-based test, two-way ANOVA yielded an F-Value of 37.74 and a P-Value of 2.16×10^{-9}, lower than the threshold alpha value of 0.05. Therefore, we replaced the null hypothesis (no difference exists in reaction time among groups given different versions of polygon-based visualization) with the alternative hypothesis (a difference exists in reaction time among groups given different versions of polygon-based visualization). Finally, we analyzed the difference in reaction time among nested groups with different polygon-visualization and obtained an F-Value of 0.056 and a P-Value of 0.814, both of which were larger than the alpha value of 0.05. Thus, we accepted the null hypothesis (a difference exists in reaction time among nested groups given different versions of polygon-visualization), forgoing the alternative hypothesis (no difference exists in reaction time among nested groups given different versions of polygon-visualization).

Based on ANOVA test results, we concluded that a significant difference satisfying the 95% confidence level existed among the groups provided with different versions of the visualization and among different questions of the test. In order to further delve into the makeup of such differences, a post-hoc analysis was conducted using a Boxplot. Results are shown in Figure 14.

The post-hoc analysis using boxplot revealed that the first group's (given the first version of the visualization) reaction time in solving questions to find complex sentiment clusters was the longest.

They spent an average of 12.01 s to complete the task. The second group (given the second version of the visualization) took 10.41 s on average to locate the sentiment cluster, showing a reduction of 1.6 s compared to the first group. The third group (given the third version of the visualization) had the shortest reaction time in problem-solving recording, with an average of 8.8 s. Based on these results, it can be concluded that using the final visualization form which combines clustered (colored) nodes, the sentiment word MDS map, and the constellation metaphor can help users better understand the visualization relative to the first two versions while minimizing the reaction time, thereby facilitating the use of visualization.

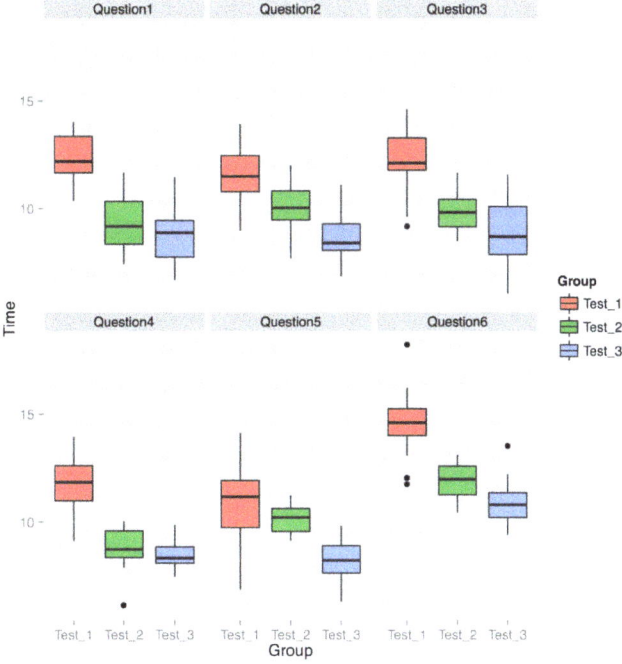

Figure 14. Post-hoc analysis for the question*group.

7. Conclusions

This study proposed three different methods for intuitive and semantic analysis based on the multilevel sentiment network visualization that we created from movie review data to serve as collective intelligence. These three methods were: (1) heatmap visualization that indicated semantic word characteristics of each node; (2) a method that described network nodes based upon a two-dimensional sentiment word map; and (3) a method using an asterism graphic for semantic interpretation of clustering followed by an evaluation to verify the suggestions.

Our system presents the insight which makes it easy to understand for users by using the three methods mentioned above.

The first evaluation revealed that participants understood the relations between locations of the nodes and the heatmap relatively well. However, their levels of awareness dropped significantly when considering nodes in the middle area of the network, which was the lowest even when subjects had been informed of relevant content. From the second evaluation, we concluded that providing heatmap visualization reinforced the understanding of emotions in each movie since it delivered emotion characteristics. The final evaluation led to the conclusion that the emotion word guideline of nodes, as well as the asterism metaphor, allowed users to understand the semantic network better than being exposed to the network solely with nodes.

Results of these three methods signify that a heatmap visualization is a useful tool for identifying subtle differences in sentiment word distribution among nodes placed in adjacent locations. This sentiment-movie network allows users to promptly understand the characteristics of the nodes since it assigns nodes based on sentiment words of each movie. Moreover, a two-dimensional distribution map showing sentiment words facilitates the understanding of the main emotion of movie nodes. Likewise, it is expected that these three methods can help users understand semantic and effective multilevel sentiment analysis in different ways.

Our results also imply that the general public could efficiently select a movie by proposing a virtual scenario for obtaining a movie recommendation using our network visualization, heatmap, and asterism graphic. This indicates that "CosMovis" could be adopted to create a novel movie recommendation algorithm that provides new contents with emotional patterns. However, this would only be possible after further improvements regarding nodes in the middle area since its awareness level needs to be enhanced according to our second evaluation.

Our research can also be applied to any numerical sentiment data (or other factors that might work as collective intelligence) as long as it can be formed within a network structure in addition to movie review data which has been covered so far. For instance, analyzing a trending issue in Twitter can help us deduce a direct and intuitive outcome if a certain issue is placed as a node in the network such as "CosMovis" by examining relations between the emotion pattern of a certain topic and those of other topics. Future research is needed to analyze data in ontology structure as well as sentiment analysis so that multilevel semantic visualization can be adopted in order to better clarify the criteria or meanings of the ontology structure data. In addition, we will conduct additional evaluation to select visual metaphorical graphics for the general public. Through this evaluation, we will select appropriate graphics which can make the general public empathize with sentiment information. Also, we will perform the three evaluations that dealt with at session 6 again in order to prove the user effectiveness considering the general public.

Author Contributions: Conceptualization, H.H.; Methodology, H.H., J.L. and K.L.; Software, H.H.; Validation, S.M.; Formal Analysis, H.H. and J.L.; Investigation, S.M.; Data Curation, H.H.; Writing—Original Draft Preparation, H.H.; Writing—Review & Editing, H.H., S.B. and K.L.; Visualization, H.H., H.H. and S.B.; Supervision, K.L.

Funding: This research was funded by [the Ministry of Education of the Republic of Korea and the National Research Foundation of Korea] grant number [NRF-2018S1A5B6075104] And [Brain Korea 21 Plus Digital Therapy Research Team] grant number [NRF31Z20130012946].

Conflicts of Interest: The authors declare no conflict of interest.

References

1. Hwansoo, L.; Dongwon, L.; Hangjung, Z. Personal Information Overload and User Resistance in the Big Data Age. *J. Intell. Inf. Syst.* **2013**, *19*, 125–139.
2. Fayyad, U.M.; Wierse, A.; Grinstein, G.G. *Information Visualization in Data Mining and Knowledge Discovery*; Morgan Kaufmann: Burlington, MA, USA, 2002.
3. Shneiderman, B.; Aris, A. Network Visualization by Semantic Substrates. *IEEE Trans. Vis. Comput. Graph.* **2006**, *12*, 733–740. [CrossRef] [PubMed]
4. Hao, M.; Rohrdantz, C.; Janetzko, H.; Dayal, U.; Keim, D.; Haug, L.E.; Hsu, M.C. Visual sentiment analysis on twitter data streams. In Proceedings of the Visual Analytics Science and Technology (VAST), Providence, RI, USA, 23–28 October 2011.
5. Thomas, M.J.; Edward, M. Graph Drawing by Force-direct Placement. *Softw. Pract. Exp.* **1991**, *21*, 1129–1164.
6. Hyoji, H.; Wonjoo, H.; Sungyun, B.; Hanmin, C.; Hyunwoo, H.; Gi-nam, K.; Kyungwon, L. CosMovis: Semantic Network Visualization by Using Sentiment Words of Movie Review Data. In Proceedings of the 19th International Conference on Information Visualisation (IV 2015), Barcelona, Spain, 21 July–24 July 2015.

7. Hyoji, H.; Gi-nam, K.; Wonjoo, H.; Hanmin, C.; Kyungwon, L. CosMovis: Analyzing semantic network of sentiment words in movie reviews. In Proceedings of the IEEE 4th Symposium on Large Data Analysis and Visualization (LDAV 2014), Paris, France, 9–10 November 2014.
8. Hyoji, H.; Hyunwoo, H.; Seongmin, M.S.; Sungyun, B.; Jihye, L.; Kyungwon, L. Visualization of movie recommendation system using the sentimental vocabulary distribution map. *J. Korea Soc. Comput. Inf.* **2016**, *21*, 19–29.
9. Doi, K.; Park, H.; Junekyu, S.; Sunyeong, P.; Kyungwon, L. Visualization of Movie Recommender System using Distribution Maps. In Proceedings of the IEEE Pacific Visualization Symposium (PacificVis 2012), Songdo, Korea, 28 February–2 March 2012.
10. Hyoji, H.; Gi-nam, K.; Kyungwon, L. A Study on Analysis of Sentiment Words in Movie Reviews and the Situation of Watching Movies. *Soc. Des. Converg.* **2013**, *43*, 17–32.
11. Deng, Z.H.; Yu, H.; Yang, Y. Identifying sentiment words using an optimization model with l1 regularization. In *Thirtieth AAAI Conference on Artificial Intelligence*; AAAI Press: Menlo Park, CA, USA, 2016.
12. MyungKyu, K.; JungHo, K.; MyungHoon, C.; Soo-Hoan, C. An Emotion Scanning System on Text Documents. *Korean J. Sci. Emot.* **2009**, *12*, 433–442.
13. JoungYeon, S.; KwangSu, C. The Perceived Lexical Space for Haptic Adjective based on Visual Texture aroused form Need for Touch. *Soc. Des. Converg.* **2013**, *38*, 117–128.
14. Yadollahi, A.; Shahraki, A.G.; Zaiane, O.R. Current state of text sentiment analysis from opinion to emotion mining. *ACM Comput. Surv. CSUR* **2017**, *50*, 25. [CrossRef]
15. Kucher, K.; Paradis, C.; Kerren, A. The state of the art in sentiment visualization. *Comput. Graph. Forum* **2018**, *37*, 71–96. [CrossRef]
16. Oard, D.W.; Marchionini, G. *A Conceptual Framework for Text Filtering*; DRUM: College Park, MD, USA, 1996; pp. 1–6.
17. Sarwar, B.; Karypis, G.; Konstan, J.; Riedl, J. Item-based Collaborative Filtering Recommendation Algorithms. In Proceedings of the 10th International World Wide Web Conference, HongKong, China, 1–5 May 2001; pp. 285–295.
18. Li, P.; Yamada, S. A Movie Recommender System Based on Inductive Learning. In Proceedings of the 2004 IEEE Conference on Cybernetics and Intelligent Systems, Singapore, 1–3 December 2004; pp. 318–323.
19. Ziani, A.; Azizi, N.; Schwab, D.; Aldwairi, M.; Chekkai, N.; Zenakhra, D.; Cheriguene, S. Recommender System Through Sentiment Analysis. In Proceedings of the 2nd International Conference on Automatic Control, Telecommunications and Signals, Annaba, Algeria, 11–12 December 2017.
20. Cody, D.; Ben, S. Motif simplification: Improving network visualization readability with fan, connector, and clique glyphs. In *Proceedings of the SIGCHI Conference on Human Factors in Computing Systems (CHI '13)*; ACM: New York, NY, USA, 2013; pp. 3247–3256.
21. Uboldi, G.; Caviglia, G.; Coleman, N.; Heymann, S.; Mantegari, G.; Ciuccarelli, P. Knot: an interface for the study of social networks in the humanities. In *Proceedings of the Biannual Conference of the Italian Chapter of SIGCHI (CHItaly '13)*; ACM: New York, NY, USA, 2013; Volume 15, pp. 1–9.
22. Henry, N.; Bezerianos, A.; Fekete, J. Improving the Readability of Clustered Social Networks using Node Duplication. *IEEE Vis. Comput. Graph.* **2008**, *14*, 1317–1324. [CrossRef] [PubMed]
23. Kang, G.J.; Ewing-Nelson, S.R.; Mackey, L.; Schlitt, J.T.; Marathe, A.; Abbas, K.M.; Swarup, S. Semantic network analysis of vaccine sentiment in online social media. *Vaccine* **2017**, *35*, 3621–3638. [CrossRef] [PubMed]
24. DougWoong, H.; HyeJa, K. Appropriateness and Frequency of Emotion Terms in Korea. *Korean J. Psychol. Gen.* **2000**, *19*, 78–98.
25. NAVER Movie. Available online: https://movie.naver.com/ (accessed on 1 February 2019).
26. Mecab-ko-lucene-analyzer. Available online: http://eunjeon.blogspot.kr (accessed on 15 February 2018).
27. Morpheme. Available online: https://en.wikipedia.org/wiki/Morpheme (accessed on 15 February 2018).
28. Srinivasa-Desikan, B. *Natural Language Processing and Computational Linguistics: A Practical Guide to Text Analysis with Python, Gensim, SpaCy, and Keras*; Packt Publishing Ltd.: Birmingham, UK, 2018.
29. Wilkinson, L.; Frendly, M. The History of the Cluster Heat Map. *Am. Stat.* **2009**, *63*, 179–184. [CrossRef]
30. Van Eck, N.J.; Waltman, L.; Den Berg, J.; Kaymak, U. Visualizing the computational intelligence field. *IEEE Comput. Intell. Mag.* **2006**, *1*, 6–10. [CrossRef]

31. Robert, G.; Nick, G.; Rose, K.; Emre, S.; Awalin, S.; Cody, D.; Ben, S. Meirav Taieb-Maimon NetVisia: Heat Map, Matrix Visualization of Dynamic Social Network Statistics & Content. In Proceedings of the Third IEEE International Conference on Social Computing (the SocialCom 2011), Boston, MA, USA, 9–11 October 2011.
32. Young-Sik, J.; Chung, Y.J.; Jae Hyo, P. Visualisation of efficiency coverage and energy consumption of sensors in wireless sensor networks using heat map. *IET Commun.* **2011**, *5*, 1129–1137.
33. Fu, Y.; Liu, R.; Liu, Y.; Lu, J. Intertrochanteric fracture visualization and analysis using a map projection technique. *Med. Biol. Eng. Comput.* **2019**, *57*, 633–642. [CrossRef]
34. Reinhardt, W.; Moi, M.; Varlem, T. Artefact-Actor-Networks as tie between social networks and artefact networks. In Proceedings of the 5th International Conference on Collaborative Computing: Networking, Applications and Worksharing, Washington, DC, USA, 11–14 November 2009.
35. Flocking Algorithms. Available online: https://en.wikipedia.org/wiki/Flocking_(behavior) (accessed on 15 February 2018).
36. Risch, J.S. On the role of metaphor in information visualization. *arXiv* **2008**, arXiv:0809.0884.
37. Hiniker, A.; Hong, S.; Kim, Y.S.; Chen, N.C.; West, J.D.; Aragon, C. Toward the operationalization of visual metaphor. *J. Assoc. Inf. Sci. Technol.* **2017**, *68*, 2338–2349. [CrossRef]
38. Wiebe, K.L.; Bortolotti, G.R. Variation in carotenoid-based color in northern flickers in a hybrid zone. *Wilson J. Ornithol.* **2002**, *114*, 393–401. [CrossRef]
39. Edgell, S.E.; Stephen, E.; Noon, S.M.; Sheila, M. Effect of violation of normality on the t test of the correlation coefficient. *Psychol. Bull.* **1984**, *95*, 579. [CrossRef]
40. Clinch, J.J.; Keselman, H.J. Parametric alternatives to the analysis of variance. *J. Educ. Stat.* **1982**, *7*, 207–214. [CrossRef]
41. Fujikoshi, Y. Two-way ANOVA models with unbalanced data. *Discret. Math.* **1993**, *116*, 315–334. [CrossRef]

© 2019 by the authors. Licensee MDPI, Basel, Switzerland. This article is an open access article distributed under the terms and conditions of the Creative Commons Attribution (CC BY) license (http://creativecommons.org/licenses/by/4.0/).

Article

Sentiment Classification Using Convolutional Neural Networks

Hannah Kim and Young-Seob Jeong *

Department of Future Convergence Technology, Soonchunhyang University, Asan-si 31538, Korea; hannah@sch.ac.kr
* Correspondence: bytecell@sch.ac.kr

Received: 29 April 2019; Accepted: 4 June 2019; Published: 7 June 2019

Abstract: As the number of textual data is exponentially increasing, it becomes more important to develop models to analyze the text data automatically. The texts may contain various labels such as gender, age, country, sentiment, and so forth. Using such labels may bring benefits to some industrial fields, so many studies of text classification have appeared. Recently, the Convolutional Neural Network (CNN) has been adopted for the task of text classification and has shown quite successful results. In this paper, we propose convolutional neural networks for the task of sentiment classification. Through experiments with three well-known datasets, we show that employing consecutive convolutional layers is effective for relatively longer texts, and our networks are better than other state-of-the-art deep learning models.

Keywords: deep learning; convolutional neural network; sentiment classification

1. Introduction

In the Big Data era, the amount of various data, such as image, video, sound, and text, is increasing exponentially. As text is the largest among them, studies related to text analysis have been actively conducted from the past to the present. In particular, text classification has drawn much attention because the text may have categorical labels such as sentiment (e.g., positive or negative), author gender, language category, or various types (e.g., spam or ham). For example, the users of Social Network Services (SNS) mostly represent their sentimental feeling, and they often share some opinions about daily news with the public or friends. Emotional analysis involves mood categories (e.g., happiness, joy, satisfaction, angry), while sentiment analysis involves categories such as positive, neutral, and negative. In this paper, we target the sentiment analysis that classifies the given text into one of sentiment categories. On websites about movies, people are likely to post their comments that probably contain sentiment or opinions. If such a sentiment is accurately predicted, then it will be applicable to various industrial fields (e.g., movie recommendation, personalized news-feed). Indeed, the international market of movies is growing much faster than before, so many companies (e.g., Netflix) provide movie recommendation services that essentially predict the sentiment or rating scores of customers.

There have been many studies that have adopted machine learning techniques for text classification. Although the machine learning techniques have been widely used and have shown quite successful performance, they strongly depend on manually-defined features, where the feature definition requires much effort of domain experts. For this reason, deep learning techniques have been drawing attention recently, as they may reduce the effort for the feature definition and achieve relatively high performance (e.g., accuracy). In this paper, we aim at sentiment classification for text data and propose an architecture of the Convolutional Neural Network (CNN) [1–3], which is a type of deep learning model. We demonstrate the effectiveness of our proposed network through experimental comparison with other machine learning models.

The contributions of this paper can be summarized as follows: (1) we design an architecture of two consecutive convolutional layers to improve performance for long and complex texts; (2) we provide a discussion about the architectural comparison between our models and two other state-of-the-art models; (3) we apply the CNN model to binary sentiment classification and achieved 80.96%, 81.4%, and 70.2% weighted F1 scores for Movie Review (MR) data [4–6], Customer Review (CR) data [7], and Stanford Sentiment Treebank (SST) data [8], respectively; and (4) we also show that our model achieved 68.31% for ternary sentiment classification with the MR data.

The remainder of this paper is organized as follows. Section 2 reviews related studies about sentiment classification, machine learning, and deep learning models. Section 3 presents a detailed description of our proposed network. Section 4 describes the experimental settings and datasets and compares the proposed model with some other models. Thereafter, Section 5 discusses the experimental results. Finally, Section 6 concludes this paper.

2. Background

2.1. Machine Learning for Sentiment Classification

The sentiment can be defined as a view of or an attitude toward a situation or event. It often involves several types of self-indulgent feelings: happiness, tenderness, sadness, or nostalgia. One may define the sentiment labels as a polarity or valence (e.g., positive, neutral, and negative) or several types of emotional feeling (e.g., angry, happy, sad, proud). The definition of sentiment labels will affect the outcome of the sentiment analysis, so we need to define the sentiment labels carefully. There have been studies that defined three or more sentiment labels (e.g., opinion rating scores, emotional feelings) [9–13], and some studies adopted two-dimensional labels (e.g., positive and negative) [14–26]. Although there has been much performance improvement in the field of sentiment analysis, the binary classification of sentiment still remains challenging; for example, the performance (e.g., accuracy) of recent studies varied from 70–90% in terms of the characteristics of data. In this paper, we aim at the binary classification (positive or negative) and the ternary classification (positive, neutral, and negative).

There have been many studies on classifying sentiments using machine learning models, such as Support Vector Machine (SVM), Naive Bayes (NB), Maximum Entropy (ME), Stochastic Gradient Descent (SGD), and ensemble. The most widely-used features for such machine learning models have been n-grams. Read [14] used unigram features for sentiment binary classification and obtained 88.94% accuracy using the SVM. Kennedy and Inkpen [15] adopted unigram and bigram features for sentiment binary classification and achieved 84.4% accuracy for movie review data [27] using the SVM. Wan [20] tackled binary classification with unigram and bigram features and achieved 86.1% accuracy for translated Amazon product review data. In [21], Akaichi utilized a combination of unigram, bigram, and trigram features and obtained 72.78% accuracy using the SVM. Valakunde and Patwardhan [28] aimed at five-class (e.g., strong positive, positive, neutral, negative, and strong negative) sentiment classification and obtained 81% accuracy using the SVM with bigram features. In the study of Gautam and Yadav [18], they utilized the SVM along with the semantic analysis model for the sentiment binary classification of Twitter texts and achieved 89.9% accuracy using unigram features. Tripathy et al. [19] also used the SVM with n-gram features and obtained 88.94% accuracy for sentiment binary classification. Hasan et al. [25] adopted unigram features for sentiment binary classification and achieved 79% using the NB for translated Urdu tweets data. All of these studies that utilized n-gram features generally achieved 70–90% accuracies, and the most effective model was the SVM.

There also have been studies that have defined hand-crafted features for sentiment classification. Yassine and Hajj [29] used affective lexicon, misspelling, and emoticons as features and obtained

87% accuracy for ternary classification using the SVM. Denecke [22] defined three scores (e.g., positivity, negativity, and objectivity) as features and achieved 67% precision and 66% recall for binary classification using the Logistic Regression (LR) classifier. Jiang et al. [16] tackled binary classification using the SVM and achieved 67.8% accuracy for Twitter texts, where they adopted two kinds of target-independent features (e.g., twitter content features and sentiment lexicon features). Bahrainian and Dengel [23] used the number of positive words and the number of negative words as features and achieved 86.7% accuracy for binary classification with Twitter texts. Neethu and Rajasree [17] combined unigram features and their own Twitter-specific features and obtained 90% accuracy for binary classification using the SVM. Karamibekr and Ghorbani [30] defined the number opinion nouns as a feature and combined the feature with unigrams. They achieved 65.46% accuracy for ternary classification using the SVM. Antai [24] used the normalized frequencies of words as features and obtained 84% accuracy for binary classification using the SVM. Ghiassi and Lee [31] aimed at five-class sentiment classification, and they defined a domain-independent feature set for Twitter data. They achieved a 92.7% F1 score using the SVM. Mensikova and Mattmann [26] utilized the results of Named Entity (NE) extraction as features and obtained a 0.9 False Positive Rate (FPR). These studies elaborated on feature definition rather than using only n-gram features, and they accomplished better performance (e.g., 92.7% F1 score). It is not fair, of course, to compare the performance between the previous studies because they used different datasets. However, as shown in [17,30], it is obvious that the combination of hand-crafted features and the n-grams must be better than using only n-gram features.

Although there has been success using machine learning models with the hand-crafted features and the n-gram features, these studies have a common limitation that their performance varied depending how well the features were defined; for different data, much effort of domain experts will be required to achieve better performance. This limitation also exists in information fusion approaches for sentiment analysis [32] that combine other resources (e.g., ontology, lexicon), as it will cost much time and effort of domain experts. Deep learning model is one of the solutions for such a limitation, as it is known to capture arbitrary patterns (i.e., features) automatically. Furthermore, as stated in [33], using the deep learning model for sentiment analysis will provide a meta-level feature representation that generalizes well on new domains.

In this paper, we take the deep learning technique to tackle the sentiment classification. In the next subsection, we review previous studies that adopted the deep learning techniques for the sentiment classification.

2.2. Deep Learning for Sentiment Classification

The deep learning technique, one of the machine learning techniques, has been recently widely used to classify sentiments. Dong et al. [34] designed a new model, namely the Adaptive Recursive Neural Network (AdaRNN), that classifies the Twitter texts into three sentiment labels: positive, neutral, and negative. By experimental results, the AdaRNN achieved 66.3% accuracy. Huang et al. [35] proposed Hierarchical Long Short-Term Memory (HLSTM) and obtained 64.1% accuracy on Weibo tweet texts. Tang et al. [36] introduced a new variant of the RNN model, the Gated Recurrent Neural Network (GRNN), which achieved 66.6% (Yelp 2013–2015 data) and 45.3% (IMDB data) accuracies. All of these studies commonly assumed that there were three or more sentiment labels.

Meanwhile, Qian et al. [37] utilized Long Short-Term Memory (LSTM) [38–41] for binary classification of sentiment and obtained 82.1% accuracy on the movie review data [27]. Kim [1] had a result of a maximum of 89.6% accuracy with seven different types of data through their CNN model with one convolutional layer. Zhang et al. [42] proposed three-way classification and obtained a maximum 94.2% accuracy with four datasets, where their best three-way model was NB-SVM. Severyn and Moschitti [43] employed a pretrained Word2Vec for their CNN model and achieved 84.79% (phrase-level) and 64.59% (message-level) accuracies with SemEval-2015 data. The CNN model used in [43] was essentially the same as the model of [1]. Deriu et al. [44] trained the CNN model that had a combination of two convolutional layers and two pooling layers to classify tweet data of four languages sentimentally and obtained a 67.79% F1 score. In the study of Ouyang et al. [45], the CNN model, which had three convolution/pooling layer pairs, was proposed, and the model outperformed other previous models including the Matrix-Vector Recursive Neural Network (MV-RNN) [46]. We conducted experiments with several of our CNN models that had different structures, and two of our models (e.g., the seventh and eighth models) were similar to the model of [44,45].

As summarized above, for the sentiment classification, there are two dominant types of deep learning technique: RNN and CNN. In this work, we propose a CNN model, the structure of which was carefully designed for effective sentiment classification. In the next subsection, we explain the advantages of the CNN in analyzing text.

2.3. Convolutional Neural Network for Text Classification

Among the existing studies using deep learning to classify texts, the CNN takes advantage of the so-called convolutional filters that automatically learn features suitable for the given task. For example, if we use the CNN for the sentiment classification, the convolutional filters may capture inherent syntactic and semantic features of sentimental expressions, as shown in [47]. It has been shown that a single convolutional layer, a combination of convolutional filters, might achieve comparable performance even without any special hyperparameter adjustment [1]. Furthermore, the CNN does not require expert knowledge about the linguistic structure of a target language [48]. Thanks to these advantages, the CNN has been successfully applied to various text analyses: semantic parsing [49], search by query [50], sentence modeling [2].

One may argue that the Recurrent Neural Network (RNN) [51] might be better for the text classification than for the CNN, as it preserves the order of the word sequence. However, the CNN is also capable of capturing sequential patterns, as concerns the local patterns by the convolutional filters; for example, the convolutional filters along with the attention technique have been successfully applied to machine translation [52]. Moreover, compared to the RNN, the CNN mostly has a smaller number of parameters, so that the CNN is trainable with a small amount of data [43]. The CNN is also known to explore the richness of pretrained word embeddings [53].

In this paper, we design a CNN model for the sentiment classification and show that our network is better than other deep learning models through experimental results.

3. The Proposed Method

CNN, which has been widely used on image datasets, extracts the significant features of the image, as the "convolutional" filter (i.e., kernel) moves through the image. If the input data are given as one-dimensional, the same function of CNN could be used in the text as well. In the text area, while the filter moves, local information of texts is stored, and important features are extracted. Therefore, using CNN for text classification is effective.

Figure 1 shows a graphical representation of the proposed network. The network consisted of an embedding layer, two convolutional layers, a pooling layer, and a fully-connected layer. We padded the sentence vectors to make a fixed size. That is, too long sentences were cut to a certain length, and too short sentences were appended with the [PAD] token. We set the fixed length S to be the maximum length of the sentences. An embedding layer that maps each word of a sentence to an E-dimensional feature vector outputs an $S \times E$ matrix, where E denotes the embedding size. For example, suppose that 10 is king, 11 is shoes, and 20 is queen in the embedding space. 10 and 20 are close in this space due to the semantic similarity of king and queen, but 10 and 11 are quite far because of the semantic dissimilarity of king and shoes. In this example, 10, 11, and 20 are not numeric values, they are just the simple position in this space. In other words, the embedding layer is a process of placing words received as input into a semantically well-designed space, where words with similar meanings are located close and words with opposite meanings are located far apart, digitizing them into a vector. The embedding is the process of projecting a two-dimensional matrix into a low-dimensional vector space (*E*-dimension) to obtain a word vector. The embedding vectors can be obtained from other resources (e.g., Word2Vec) or from the training process. In this paper, our embedding layer was obtained through the training process, and all word tokens including the [UNK] token for unseen words would be converted to numeric values using the embedding layer.

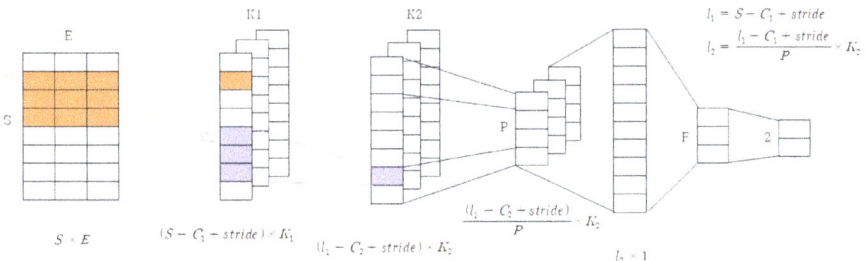

Figure 1. The graphical representation of the network, where the output dimensions of each layer are represented at the bottom of the corresponding layers.

The $S \times E$ matrix, the output of the embedding layer, is then laid down as the first convolutional layer. The first convolutional layer is the $C_1 \times E$ matrix, which stores the local information needed to classify the sentiment class in a $S \times E$ matrix and convey information to the next convolutional layer. The $C_1 \times E$ matrix slides (i.e., convolves) all the values of the $S \times E$ matrix with an arbitrary stride, calculates the dot product, and passes the dot product result to the next layer. The second convolutional layer uses the $C_2 \times 1$ matrix to extract features from the contextual information of the main word based on the local information stored in the first convolutional layer. C_1 and C_2 denote the filter size of each convolutional layer, and the two convolutional layers have K_1 and K_2 distinct filters, respectively, to capture unique contextual information. In other words, the first convolutional layer is utilized to look at simple contextual information while looking over the $S \times E$ matrix, and the second convolutional layer is utilized to capture key features and then extract them (e.g., worst, great) that contain sentiments affecting classification.

The matrix that passed through the consecutive convolutional layer is used as the input to the pooling layer. While average-pooling and L2-norm pooling have been used as the pooling layer position, in this paper, we used the max-pooling, which is a technique for selecting the largest value as a representative of the peripheral values. Since the sentiment is often determined by a combination of several words rather than expressing the sentiment in every word in the sentence, we adopted the max-pooling technique. The pooling layer slides all the values of the matrix, which is the output of the second convolutional layer, with an arbitrary stride, resulting in output vectors. Since max-pooling is the layer that passes to the next layer the largest value among several values, it results in output vectors of a much smaller size. In other words, the convolutional layer looks at the context and extracts the main features, and the pooling layer plays a role in selecting the most prominent features.

After passing through the pooling layer, a flattening process is performed to convert the two-dimensional feature map from the output into a one-dimensional format and to delivery it to an *F*-dimensional Fully-Connected (FC) layer. Since the FC layer takes a one-dimensional vector as the input, the two-dimensional vector delivered from the pooling layer needs to be flattened. The FC layer connects all input and output neurons. A vector that passes through the FC layer forms an output that is classified as positive or negative. The activation function softmax functions to classify multiple classes in the FC layer. The softmax function outputs, the value, which is the probability value, is generated for each class.

Most people might think that with many convolutional layer stacks, it may be better to store local information and to extract contextual information; however, deep networks do not always have higher performance than shallow networks. Because a network's variation (e.g., deep or shallow) might rely on the length of the data and the number of data and features, in this paper, we argue that data passing through two consecutive convolutional layers and then passing through the pooling layer is successful at storing context information and extracting prominent features. This assertion is demonstrated by experiments and discussed in Section 5.

4. Experiment

4.1. Data

We used three Movie Review (MR) data from kaggle [4–6]. The dataset in [4] consisted of two labels, positive and negative, while [5] was composed of three labels of positive, neutral, and negative. Furthermore, the dataset in [6] was composed of five labels of positive, somewhat positive, neutral, somewhat negative, and negative. The positive class of [4], the positive class of [5], and the positive and somewhat positive classes of [6] were merged and labeled as the positive class. The neutral class of [5] and the neutral class of [6] were merged and labeled as the neutral class. Then, the negative class of [4], the negative class of [5], and the somewhat negative and negative classes of [6] were merged and labeled as the negative class. Positive, neutral, and negative were assigned as 1, 0, and 2, respectively. Some 11,800 positive data, 5937 neutral data, and 9698 negative data were used, totaling 27,435. Amazon Customer Review data (CR) [7] were labeled as positive and negative, and the amount of positive data was more than that of negative data. Therefore, we used only 3671 out of 34,062 data to control the positive:negative ratio. Stanford Sentiment Treebank data (SST) [8] were labeled as a value between 0 and 1, so a value of 0.5 or more was relabeled as positive, and a value of less than 0.5 was relabeled as negative.

Table 1 shows the detailed statistics of the data. N is the number of text data, Dist (+, −) is the ratio of positive:negative classes, and aveL/maxL is the average length/maximum length of the text. We divide all the data into three sets (i.e., train, validation, and test) with a ratio of 55:20:25, and $|V|$ was the dictionary size of each dataset.

Table 1. Detailed statistics of the data.

Data	N	Dist (+,−)	aveL/maxL	Train:Test:Val	$|V|$
MR	21,498	55:45	31/290	12,095:5375:4031	9396
CR	3671	62:38	19/227	2064:918:689	1417
SST	11,286	52:48	12/41	6348:2822:2116	3550

4.2. Preprocessing

Preprocessing was carried out to modify the text data appropriately in the experiment. We used decapitalization and did not mark the start and end of the sentences. We deleted #, two or more spaces, tabs, Retweets (RT), and stop words. We also changed the text that represented the url that began with "http" to [URL] and the text that represented the account ID that began with "@" to [NAME]. In addition, we changed digits to [NUM], and special characters to [SPE]. We changed "can't" and "isn't" to "can not" and "is not", respectively, since "not" is important in sentiment analysis.

We split the text data by space and constructed a new dictionary using only the words appearing more than six times in the whole text. In the MR data, 9394 words out of a total of 38,195 words were made into a dictionary. In CR data, 1415 out of a total of 5356 words, and in SST data, 3548 out of a total of 16,870 words were made. The new dictionary also included [PAD] for padding and [UNK] to cover missing words. Padding is a method of cropping all text to a fixed length and filling it with the [PAD] token for text shorter than that length. Here, the fixed length was the maximum length of text in each dataset.

4.3. Performance Comparison

Decision Tree (DT), Naive Bayes (NB), Support Vector Machine (SVM) [54], and Random Forest (RF) [55] were used for comparing with our CNN models. Table 2 summarizes the description of the traditional machine learning models and the parameter settings. The parameters were optimized via a grid searching. We also compared our CNN models with the state-of-the-art models such as Kim's [1] and Zhang et al.'s [42].

Table 2. The description of the traditional machine learning models and the parameter settings.

Model	Description
Naive Bayes (NB)	- Probabilistic classifier based on the Bayes' theorem - Assumes the independence between features
Decision Tree (DT)	- C4.5 classifier using the J48 algorithm - Confidence factor for pruning: 0.25 - Minimum number of instances: 1 - The number of decimal places: 1 - Determines the amount of data used for pruning: 5 - Reduced error pruning: True
Support Vector Machine (SVM)	- Non-probabilistic binary classification model that finds a decision boundary with a maximum distance between two classes - Kernel: RBF - Exponent = 1.0, Complexity (c) = 10.0
Random Forest (RF)	- Kind of ensemble model that generates final result by incorporating the results of multiple decision trees - Maximum tree depth: 30 - Number of iterations: 150 - Each tree has no depth limitation

In order to optimize the CNN model proposed in this paper, we experimented with various settings, and the optimal settings were as follows: (1) the embedding dimension $E = 25$; (2) C_1 and C_2 were set to 3; (3) K_1 and K_2 were 16 and 8; and (4) the pooling layer dimension $P = 2$. We set the strides for the kernels to be 1. Every layer took the Rectified Linear Unit (ReLU) as an activation function except for the output layer that took a softmax function. The number of the trainable parameter was 476,331, and we adopted Adam's optimizer with an initial learning rate of 0.001. We took the L2 regularization for the fully-connected layer. The number of epochs was changed according to the complexity of the network (i.e., the number of layers and parameters). For example, the first network was trained for 5 epochs, and the model of Kim [1] was trained for 10 epochs.

In the study of Zhang et al. [42], NB-SVM was the best. The effectiveness of the combination with different models (e.g., NB, SVM, NB-SVM) may vary for different tasks. We tested different combinations with our datasets and discovered that the SVM with the RBF kernel was the best enhanced model for our task. Thus, the CNN enhanced by SVM was used for comparison with other models.

5. Result and Discussion

5.1. Result

The three datasets (movie review data, customer review data, and Stanford Sentiment Treebank data) were applied to the proposed CNN models, traditional machine-learning models, and other state-of-the-art models. Note that, we conducted two experiments: binary classification and ternary classification. Tables 3 and 4 show the experimental results of binary classification, where the two values for each cell correspond to the "positive" and "negative" classes, respectively. Table 5 shows the experimental results of ternary classification in MR data. Table 3 shows the experimental results with the MR data, while Table 4 describes the experimental results with the CR and SST data. These tables include the accuracy, precision, recall, F1 score, and weighted-F1 score. For example, the F1 scores of the decision tree are 67.2% for positive and 31.1% for negative in Table 3.

Table 3. The results with the MR data, where Emb, Conv, Pool, globalpool, FC stand for embedding layer, convolutional layer, pooling layer, global pooling layer, and fully-connected layer, respectively.

Model	Accuracy	Precision	Recall	F1	Weighted-F1
Decision Tree	59.64	58.0/64.0	76.2/39.6	67.4/47.1	57.2
Naive Bayes	56.40	57.5/53.3	77.7/30.7	66.1/38.9	52.0
Support Vector Machine	54.95	57.7/55.2	93.0/9.1	69.3/15.5	44.9
Random Forest	58.73	56.4/59.8	74.7/39.4	66.4/46.4	58.1
Kim [1]	80.85	80.7/80.9	76.2/84.7	78.3/82.8	80.75
Zhang et al. [42]	77.28	72.4/69.1	56.1/82.1	63.2/75.0	69.62
Emb+Conv+Conv+Pool+FC	81.06	81.5/80.7	75.6/85.6	78.4/83.1	80.96
Emb+Conv+Pool+FC	79.70	77.4/81.63	78.2/80.9	77.8/81.3	79.71
Emb+Conv+Conv+Conv+Pool+FC	80.30	80.2/80.4	75.3/84.5	77.7/82.4	80.26
Emb+Conv+Pool+Conv+FC	78.17	74.4/81.8	79.5/77.1	76.8/79.4	78.22
Emb+Conv+globalpool+FC	77.54	77.3/77.7	71.8/82.4	74.4/79.9	77.39
Emb+Conv+Conv+globalpool+FC	79.06	79.1/79.0	73.5/83.8	76.2/81.3	78.98
Emb+Conv+Pool+Conv+Pool+FC	79.11	78.6/79.5	74.3/83.1	76.4/81.2	79.0
Emb+Conv+Pool+Conv+Pool+Conv+Pool+FC	74.61	84.1/72.8	59.5/90.6	69.7/80.7	75.7

Table 4. The results with the CR and SST data, where Emb, Conv, Pool, globalpool, FC stand for embedding layer, convolutional layer, pooling layer, global pooling layer, and fully-connected layer, respectively.

Model	CR		SST	
	Weighted-F1	F1	Weighted-F1	F1
Decision Tree	63.7	47.0/74.5	51.5	62.4/41.7
Naive Bayes	61.0	77.7/31.8	35.7	10.6/63.4
Support Vector Machine	59.7	26.5/78.5	37.8	68.6/4.0
Random Forest	64.4	41.8/76.9	51.2	59.0/47.3
Kim [1]	74.8	78.6/65.6	56.1	47.2/66.3
Zhang et al. [42]	54.8	64.7/37.7	52.1	45.6/59.5
Emb+Conv+Conv+Pool+FC	78.3	84.8/67.1	68.3	70.5/65.7
Emb+Conv+Pool+FC	78.3	82.3/71.3	66.5	67.7/65.1
Emb+Conv+Conv+Conv+Pool+FC	70.4	82.0/50.3	68.2	68.4/68.0
Emb+Conv+Pool+Conv+FC	75.3	84.0/60.3	68.6	70.5/66.5
Emb+Conv+globalpool+FC	81.4	86.1/73.4	70.2	72.8/67.2
Emb+Conv+Conv+globalpool+FC	79.4	84.5/70.5	70.0	71.2/68.7
Emb+Conv+Pool+Conv+Pool+FC	73.18	82.8/56.5	66.62	67.7/65.4
Emb+Conv+Pool+Conv+Pool+Conv+Pool+FC	51.57	77.6/6.7	65.21	70.2/59.5

In Table 3, the Random Forest (RF) achieved the best F1 scores, while the Decision Tree (DT) exhibited comparable results. Our first network was the best among all models including the state-of-the-art models of [1,42]. Similarly, in Table 4, the RF and the DT achieved the best F1 scores among the traditional models, but our fifth network outperformed all other models. Note that the best networks in Tables 3 and 4 were different. That is, the first network was the best with the MR data, which had relatively longer sentences, as shown in Table 3, whereas the fifth network was the best for the other two datasets (e.g., CR, SST) that had relatively shorter sentences.

One may argue that stacking more convolutional layers might be better, as the deeper network is known to capture higher level patterns. This might be true, but it should be noted that the deeper network is not always better than the shallow networks. The depth of the network needs to be determined according to the data characteristics; too deep networks will probably overfit, whereas too shallow networks will underfit. For example, although the third network was deeper than the first network, the first network outperformed the third network, as shown in Table 3. To clarify this, we conducted experiments varying the number of convolutional layers from 1–5, and its result is depicted in Figure 2. Meanwhile, the first network can be compared with the fourth network to show that stacking two consecutive convolutional layers is better. The fifth and sixth networks may be compared with the first and second networks, if one can see if the max-pooling and the global max-pooling contribute to the models. The seventh and eighth networks had similar structures to that in [44,45].

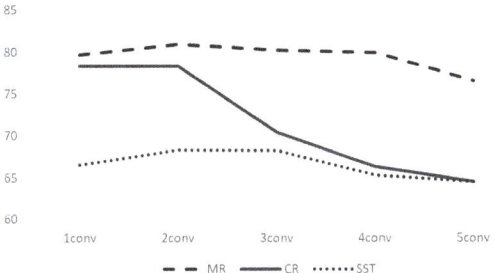

Figure 2. Classification performance according to the depth of the convolutional layer, where the horizontal axis means the number of stacked convolutional layers.

We also conducted ternary classification. The binary classification assumed that there were only two labels (e.g., positive and negative), whereas the ternary classification assumed that there were three labels (e.g., positive, negative, and neutral). As shown in Table 5, our first network was the best for the ternary classification, but the performance fell to a 68.31% weighted-F1 score.

Table 5. Performance of ternary classification with the MR data, where Emb, Conv, Pool, globalpool, FC stand for embedding layer, convolutional layer, pooling layer, global pooling layer, and fully-connected layer, respectively.

Model	Weighted-F1
Decision Tree	46.8
Naive Bayes	27.2
Support Vector Machine	34.9
Random Forest	46.2
Kim [1]	68.2
Zhang et al. [42]	67.5
Emb+Conv+Conv+Pool+FC	68.3
Emb+Conv+Pool+FC	55.0
Emb+Conv+Conv+Conv+Pool+FC	64.6
Emb+Conv+Pool+Conv+FC	66.9
Emb+Conv+globalpool+FC	65.3
Emb+Conv+Conv+globalpool+FC	67.5

5.2. Discussion

5.2.1. Comparison with Other Models

In terms of the F1 scores, our network was about 10% greater than the traditional machine models. We believe that the biggest reason is the inherent ability of the CNN to capture local patterns. The CNN has the ability to capture local patterns and higher level patterns through its convolutional and pooling layers, as explained in [48]. We believe that such an ability of the CNN model produces the performance gap. The proposed network was also better than the state-of-the-art deep learning models such as those of Kim [1] and Zhang et al. [42]. Kim [1] and Zhang et al. [42] commonly utilized three filters (sizes of 3, 4, and 5) in parallel, in the hope to capture the filters' arbitrary features with different sizes. On the other hand, our network had consecutive filters that had two benefits: (1) hierarchical (higher) features were extracted, and (2) the high-level filters would have a broader range to see local patterns, where such advantages of consecutive filters were already reported in many previous studies [56–58]. The three-way approach, employed in [42], gave worse performance than the model of [1], although it was designed to combine of a CNN model and a traditional machine learning model. The reason can be explained by the fact that the CNN model itself was already sufficient to achieve the high performance, as shown in Table 3, so the combination with other traditional models (e.g., Naive Bayes) degraded the overall performance. In Zhang et al. [42], among the sentences classified by CNN [1], the sentences with weak confidence were re-classified by the traditional machine learning models, and it improved the overall performance. On the other hand, for the three datasets in this paper, we found that the ratio of weakly-classified instances was biased, as shown in Figure 3; this means that the entropy of the weakly-classified instances was small, so it was difficult to expect that the performance would improve by the combination with the traditional models.

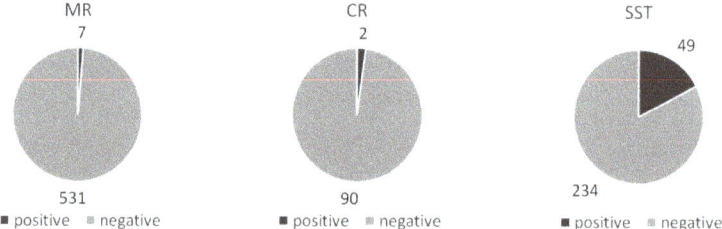

Figure 3. The ratio of positive/negative sentences with weak confidence.

5.2.2. Network Structure

As shown in Tables 3 and 4, the max-pooling had better performance in MR data, whereas the global max-pooling seemed helpful in the CR data and the SST data. The max-pooling layer gave the largest value in a certain subarea as an output, while the global max-pooling did this in the whole area. Figure 4 shows the difference.

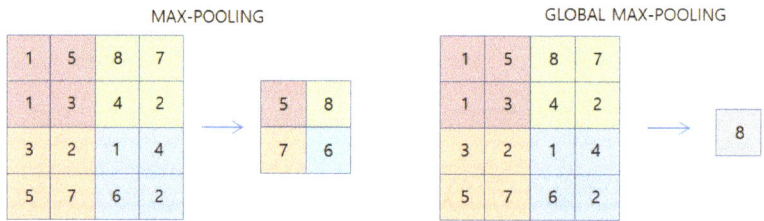

Figure 4. The difference of max-pooling and global max-pooling.

If MR data adopted global max-pooling, the large number of missing values would be discovered. Because the length of the total sentence of the MR data was longer than that of CR data and SST data, the performance of the MR data was lower. On the other hand, CR data and SST data with relatively a short length of sentences were not degraded even when using global max-pooling and, in some cases, performed better than using max-pooling. As a result, CR data and SST data showed the highest performance when one convolutional layer and global max-pooling layer were combined. The reason is also that they were shorter and simpler data than the MR data, so they showed higher performance in relatively monotonous models. However, The network with two convolutional layers never lagged behind the traditional machine learning models and state-of-the-art models in terms of performance in this experimental result.

6. Conclusions

In this paper, we designed a convolutional neural network for the sentiment classification. By experimental results, we showed that the consecutive convolutional layers contributed to better performance on relatively long text. The proposed CNN models achieved about 81% and 68% accuracies for the binary classification and ternary classification, respectively. As future work, we will apply our work to other classification tasks (e.g., gender classification). We will also continue finding better structures for the sentiment classification; for example, residual connection for stacking more layers.

Author Contributions: Conceptualization, H.K.; validation, H.K., Y.-S.J.; investigation, H.K.; resources, Y.-S.J.; data curation, H.K.; writing–original draft preparation, H.K.; writing–review and editing, Y.-S.J.; visualization, H.K.; supervision, Y.-S.J.; project administration, Y.-S.J.

Acknowledgments: This work was supported by the National Research Foundation of Korea(NRF) grant funded by the Korea government(MSIP; Ministry of Science, ICT & Future Planning) (No. 2019021348). This work was supported by the Soonchunhyang University Research Fund.

Conflicts of Interest: The authors declare no conflict of interest.

References

1. Kim, Y. Convolutional neural networks for sentence classification. *arXiv* **2014**, arXiv:1408.5882.
2. Nal, K.; Grefenstette, E.; Blunsom, P. A convolutional neural network for modelling sentences. *arXiv* **2014**, arXiv:1404.2188.
3. Lei, T.; Barzilay, R.; Jaakkola, T. Molding cnns for text: Non-linear, non-consecutive convolutions. *arXiv* **2015**, arXiv:1508.04112.
4. Amazon Movie Review Dataset. Available online: https://www.kaggle.com/ranjan6806/corpus2#corpus/ (accessed on 11 November 2012).
5. Movie Review Dataset. Available online: https://www.kaggle.com/ayanmaity/movie-review#train.tsv/ (accessed on 11 November 2012).
6. Rotten Tomatoes Movie Review Dataset. Available online: https://www.kaggle.com/c/movie-review-sentiment-analysis-kernels-only/ (accessed on 11 November 2012).
7. Consumer Reviews Of Amazon Products Dataset. Available online: https://www.kaggle.com/datafiniti/consumer-reviews-of-amazon-products/ (accessed on 11 November 2012).
8. Stanford Sentiment Treebank Dataset. Available online: https://nlp.stanford.edu/sentiment/code.html/ (accessed on 11 November 2012).
9. Pak, A.; Paroubek, P. Twitter as a corpus for sentiment analysis and opinion mining. *LREc* **2010**, *10*, 1320–1326.
10. Alm, C.O.; Roth, D.; Sproat, R. Emotions from text: Machine learning for text-based emotion prediction. In Proceedings of the Conference on Human Language Technology and Empirical Methods in Natural Language Processing, Vancouver, BC, Canada, 6–8 October 2005; pp. 579–586.
11. Bartlett, M.S.; Littlewort, G.; Frank, M.; Lainscsek, C.; Fasel, I.; Movellan, J. Recognizing facial expression: Machine learning and application to spontaneous behavior. In Proceedings of the 2005 IEEE Computer Society Conference on Computer Vision and Pattern Recognition (CVPR'05), San Diego, CA, USA, 20–25 June 2005; Volume 2, pp. 568–573.
12. Heraz, A.; Razaki, R.; Frasson, C. Using machine learning to predict learner emotional state from brainwaves. In Proceedings of the Seventh IEEE International Conference on Advanced Learning Technologies (ICALT 2007), Niigata, Japan, 18–20 July 2007; pp. 853–857.
13. Yujiao, L.; Fleyeh, H. Twitter Sentiment Analysis of New IKEA Stores Using Machine Learning. In Proceedings of the International Conference on Computer and Applications, Beirut, Lebanon, 25–26 July 2018.

14. Read, J. Using emoticons to reduce dependency in machine learning techniques for sentiment classification. In Proceedings of the ACL Student Research Workshop, Ann Arbor, Michigan, 27 June 2005; pp. 43–48.
15. Kennedy, A.; Inkpen, D. Sentiment classification of movie reviews using contextual valence shifters. *Comput. Intell.* **2006**, *22*, 110–125. [CrossRef]
16. Jiang, L.; Yu, M.; Zhou, M.; Liu, X.; Zhao, T. Target-dependent twitter sentiment classification. In Proceedings of the 49th Annual Meeting of the Association for Computational Linguistics: Human Language Technologies, Portland, Oregon, 19–24 June 2011; Volume 1, pp. 151–160.
17. Neethu, M.; Rajasree, R. Sentiment analysis in twitter using machine learning techniques. In Proceedings of the 2013 Fourth International Conference on Computing, Communications and Networking Technologies (ICCCNT), Tiruchengode, India, 4–6 July 2013; pp. 1–5.
18. Gautam, G.; Yadav, D. Sentiment analysis of twitter data using machine learning approaches and semantic analysis. In Proceedings of the 2014 Seventh International Conference on Contemporary Computing (IC3), Noida, India, 7–9 August 2014; pp. 437–442.
19. Tripathy, A.; Agrawal, A.; Rath, S.K. Classification of sentiment reviews using n-gram machine learning approach. *Expert Syst. Appl.* **2016**, *57*, 117–126. [CrossRef]
20. Wan, X. A comparative study of cross-lingual sentiment classification. In Proceedings of the 2012 IEEE/WIC/ACM International Joint Conferences on Web Intelligence and Intelligent Agent Technology, Macau, China, 4–7 December 2012; Volume 1, pp. 24–31.
21. Akaichi, J. Social networks' Facebook'statutes updates mining for sentiment classification. In Proceedings of the 2013 International Conference on Social Computing, Alexandria, VA, USA, 8–14 September 2013; pp. 886–891.
22. Denecke, K. Using sentiwordnet for multilingual sentiment analysis. In Proceedings of the 2008 IEEE 24th International Conference on Data Engineering Workshop, Cancun, Mexico, 7–12 April 2008; pp. 507–512.
23. Bahrainian, S.A.; Dengel, A. Sentiment analysis using sentiment features. In Proceedings of the 2013 IEEE/WIC/ACM International Joint Conferences on Web Intelligence (WI) and Intelligent Agent Technologies (IAT), Atlanta, GA, USA, 17–20 November 2013; Volume 3; pp. 26–29.
24. Antai, R. Sentiment classification using summaries: A comparative investigation of lexical and statistical approaches. In Proceedings of the 2014 6th Computer Science and Electronic Engineering Conference (CEEC), Colchester, UK, 25–26 September 2014; pp. 154–159.
25. Hasan, A.; Moin, S.; Karim, A.; Shamshirband, S. Machine learning-based sentiment analysis for twitter accounts. *Math. Comput. Appl.* **2018**, *23*, 11. [CrossRef]
26. Mensikova, A.; Mattmann, C.A. Ensemble sentiment analysis to identify human trafficking in web data. Workshop on Graph Techniques for Adversarial Activity Analytics (GTA 2018), Marina Del Rey, CA, USA, 5–9 February 2018
27. Pang, B.; Lee, L. A sentimental education: Sentiment analysis using subjectivity summarization based on minimum cuts. In Proceedings of the 42nd Annual Meeting on Association for Computational Linguistics, Barcelona, Spain, 21–26 July 2004; pp. 271–278.
28. Valakunde, N.; Patwardhan, M. Multi-aspect and multi-class based document sentiment analysis of educational data catering accreditation process. In Proceedings of the 2013 International Conference on Cloud & Ubiquitous Computing & Emerging Technologies, Pune, India, 15–16 November 2013; pp. 188–192.
29. Yassine, M.; Hajj, H. A framework for emotion mining from text in online social networks. In Proceedings of the 2010 IEEE International Conference on Data Mining Workshops, Sydney, Australia, 13 December 2010; pp. 1136–1142.
30. Karamibekr, M.; Ghorbani, A.A. A structure for opinion in social domains. In Proceedings of the 2013 International Conference on Social Computing, Alexandria, VA, USA, 8–14 September 2013; pp. 264–271.
31. Ghiassi, M.; Lee, S. A domain transferable lexicon set for Twitter sentiment analysis using a supervised machine learning approach. *Expert Syst. Appl.* **2018**, *106*, 197–216. [CrossRef]
32. Balazs, J.A.; Velásquez, J.D. Opinion mining and information fusion: A survey. *Inf. Fusion* **2016**, *27*, 95–110. [CrossRef]
33. Chaturvedi, I.; Cambria, E.; Welsch, R.E.; Herrera, F. Distinguishing between facts and opinions for sentiment analysis: Survey and challenges. *Inf. Fusion* **2018**, *44*, 65–77. [CrossRef]

34. Dong, L.; Wei, F.; Tan, C.; Tang, D.; Zhou, M.; Xu, K. Adaptive recursive neural network for target-dependent twitter sentiment classification. In Proceedings of the 52nd Annual Meeting of the Association for Computational Linguistics (Volume 2: Short Papers), Baltimore, Maryland, USA, 23–25 June 2014; Volume 2, pp. 49–54.
35. Huang, M.; Cao, Y.; Dong, C. Modeling rich contexts for sentiment classification with lstm. *arXiv* **2016**, arXiv:1605.01478.
36. Tang, D.; Qin, B.; Liu, T. Document modeling with gated recurrent neural network for sentiment classification. In Proceedings of the 2015 Conference on Empirical Methods in Natural Language Processing, Lisbon, Portugal, 17–21 September 2015; pp. 1422–1432.
37. Qian, Q.; Huang, M.; Lei, J.; Zhu, X. Linguistically regularized lstms for sentiment classification. *arXiv* **2016**, arXiv:1611.03949.
38. Mikolov, T. Statistical language models based on neural networks. Ph.D. Thesis, Brno University of Technology, Brno, Czechia, 2012.
39. Chung, J.; Gulcehre, C.; Cho, K.; Bengio, Y. Empirical evaluation of gated recurrent neural networks on sequence modeling. *arXiv* **2014**, arXiv:1412.3555.
40. Tai, K.S.; Socher, R.; Manning, C.D. Improved semantic representations from tree-structured long short-term memory networks. *arXiv* **2015**, arXiv:1503.00075.
41. Wang, F.; Zhang, Z.; Lan, M. Ecnu at semeval-2016 task 7: An enhanced supervised learning method for lexicon sentiment intensity ranking. In Proceedings of the 10th International Workshop on Semantic Evaluation (SemEval-2016), San Diego, CA, USA, 16–17 June 2016; pp. 491–496.
42. Zhang, Y.; Zhang, Z.; Miao, D.; Wang, J. Three-way enhanced convolutional neural networks for sentence-level sentiment classification. *Inf. Sci.* **2019**, *477*, 55–64. [CrossRef]
43. Severyn, A.; Moschitti, A. Unitn: Training deep convolutional neural network for twitter sentiment classification. In Proceedings of the 9th International Workshop On Semantic Evaluation (SemEval 2015), Denver, CO, USA, 4–5 June 2015; pp. 464–469.
44. Deriu, J.; Lucchi, A.; De Luca, V.; Severyn, A.; Müller, S.; Cieliebak, M.; Hofmann, T.; Jaggi, M. Leveraging large amounts of weakly supervised data for multi-language sentiment classification. In Proceedings of the 26th International Conference on World Wide Web, Perth, Australia, 3–7 April 2017; pp. 1045–1052.
45. Ouyang, X.; Zhou, P.; Li, C.H.; Liu, L. Sentiment analysis using convolutional neural network. In Proceedings of the 2015 IEEE International Conference on Computer and Information Technology; Ubiquitous Computing and Communications; Dependable, Autonomic and Secure Computing; Pervasive Intelligence and Computing, Liverpool, UK, 26–28 October 2015; pp. 2359–2364.
46. Socher, R.; Huval, B.; Manning, C.D.; Ng, A.Y. Semantic compositionality through recursive matrix-vector spaces. In Proceedings of the 2012 Joint Conference on Empirical Methods in Natural Language Processing and Computational Natural Language Learning, Jeju Island, Korea, 12–14 July 2012; pp. 1201–1211.
47. Rios, A.; Kavuluru, R. Convolutional neural networks for biomedical text classification: Application in indexing biomedical articles. In Proceedings of the 6th ACM Conference on Bioinformatics, Computational Biology and Health Informatics, Atlanta, GA, USA, 9–12 September 2015; pp. 258–267.
48. Zhang, X.; Zhao, J.; LeCun, Y. Character-level convolutional networks for text classification. In *Advances in Neural Information Processing Systems 28*; Neural Information Processing Systems Foundation, Inc.: Montreal, QC, Canada, 2015; pp. 649–657.
49. Yih, W.T.; He, X.; Meek, C. Semantic Parsing for Single-Relation Question Answering. In Proceedings of the 52nd Annual Meeting of the Association for Computational Linguistics, Baltimore, MD, USA, 23–25 June 2014; Volume 2, pp. 643–648.
50. Shen, Y.; Xiaodong, H.; Gao, J.; Deng, L.; Mensnil, G. Learning Semantic Representations Using Convolutional Neural Networks for Web Search. In Proceedings of the 23rd International Conference on World Wide Web, Seoul, Korea, 7–11 April 2014; pp. 373–374.
51. Lai, S.; Xu, L.; Liu, K.; Zhao, J. Recurrent convolutional neural networks for text classification. In Proceedings of the Twenty-Ninth AAAI Conference on Artificial Intelligence, Austin, TX, USA, 25–30 January 2015.
52. Gehring, J.; Auli, M.; Grangier, D.; Dauphin, Y.N. A convolutional encoder model for neural machine translation. *arXiv* **2016**, arXiv:1611.02344.

53. Dos Santos, C.; Gatti, M. Deep convolutional neural networks for sentiment analysis of short texts. In Proceedings of the COLING 2014, the 25th International Conference on Computational Linguistics: Technical Papers, Dublin, Ireland, 23–29 August 2014; pp. 69–78.
54. Boser, B.E.; Guyon, I.M.; Vapnik, V.N. A training algorithm for optimal margin classifiers. In Proceedings of the Fifth Annual Workshop on Computational Learning Theory, Pittsburgh, PA, USA, 27–29 July 1992; pp. 144–152.
55. Breiman, L. Random forests. *Mach. Learn.* **2001**, *45*, 5–32. [CrossRef]
56. Szegedy, C.; Liu, W.; Jia, Y.; Sermanet, P.; Reed, S.; Anguelov, D.; Erhan, D.; Vanhoucke, V.; Rabinovich, A. Going deeper with convolutions. In Proceedings of the 2015 IEEE Conference on Computer Vision and Pattern Recognition, Boston, MA, USA, 7–12 June 2015; pp. 1–9.
57. Szegedy, C.; Vanhoucke, V.; Ioffe, S.; Shlens, J.; Wojna, Z. Rethinking the inception architecture for computer vision. In Proceedings of the IEEE Conference on Computer Vision and Pattern Recognition, Las Vegas, NV, USA, 27–30 June 2016; pp. 2818–2826.
58. Szegedy, C.; Ioffe, S.; Vanhoucke, V.; Alemi, A.A. Inception-v4, inception-resnet and the impact of residual connections on learning. In Proceedings of the Thirty-First AAAI Conference on Artificial Intelligence, San Francisco, CA, USA, 4–9 February 2017.

© 2019 by the authors. Licensee MDPI, Basel, Switzerland. This article is an open access article distributed under the terms and conditions of the Creative Commons Attribution (CC BY) license (http://creativecommons.org/licenses/by/4.0/).

Article

Sentiment-Aware Word Embedding for Emotion Classification

Xingliang Mao [1,†,‡], Shuai Chang [2,‡], Jinjing Shi [2,*], Fangfang Li [2,*] and Ronghua Shi [2]

[1] Science and Technology on Information Systems Engineering Laboratory, National University of Defense Technology, Changsha 410073, China; gisor@163.com
[2] School of Computer Science and Engineering, Central South University, Changsha 410073, China; 164712054@csu.edu.cn (S.C.); shirh@csu.edu.cn (R.S.)
* Correspondence: 214009@csu.edu.cn (J.S.); lifangfang@csu.edu.cn (F.L.)
† Current address: National University of Defense Technology, Changsha 410073, China.
‡ These authors contributed equally to this work.

Received: 12 February 2019; Accepted: 22 March 2019; Published: 29 March 2019

Abstract: Word embeddings are effective intermediate representations for capturing semantic regularities between words in natural language processing (NLP) tasks. We propose sentiment-aware word embedding for emotional classification, which consists of integrating sentiment evidence within the emotional embedding component of a term vector. We take advantage of the multiple types of emotional knowledge, just as the existing emotional lexicon, to build emotional word vectors to represent emotional information. Then the emotional word vector is combined with the traditional word embedding to construct the hybrid representation, which contains semantic and emotional information as the inputs of the emotion classification experiments. Our method maintains the interpretability of word embeddings, and leverages external emotional information in addition to input text sequences. Extensive results on several machine learning models show that the proposed methods can improve the accuracy of emotion classification tasks.

Keywords: emotion classification; sentiment lexicon; text feature representation; hybrid vectorization; sentiment-aware word embedding

1. Introduction

With the rapid increase in the popularity of social media applications, such as Twitter, a larger amount of sentiment data is being generated. Emotional analysis has attracted much attention. At the same time, sentiment analysis for Chinese social network data has been gradually developed. In 2013, the second CCF (China Computer Federation) International Conference on Natural Language Processing and Chinese Computing (NLPCC) established the task of evaluating the emotions of Weibo, which attracted many researchers and institutions. The conference drove the development of emotional analysis in China. Weibo sites have been the main communication tool. They provide information that is more up-to-date than conventional news sources, and this has encouraged researchers to analyze emotional information from this data source. There are many differences between Weibo text and traditional long text, such as movie reviews in sentiment analysis. Firstly, they are short with no more than 140 characters. Secondly, words used in Weibo are more casual than those in official texts, and they contain a lot of noise, such as informal text snippets. For example, there are web-popular words, like "LanShouXiangGu" (a network language/buzzword, means feel awful and want to cry). Web-popular words might be seen as traditional words but represent different meanings or emotions. Finally, Chinese is largely different from English; it has more complex syntaxes and sentence structures. This increases the difficulty of emotional analysis in Chinese.

Text consists of many ordered words in the emotional analysis task. In order to process a text document mathematically, the text document is projected into the vector space. A text document is represented by a vector in the same vector space so that the document can be classified by a model. The BoW (bag-of-words) model is widely used in text processing applications. It is an approach to modeling texts numerically [1]. It processes texts regardless of word order and semantic structure and disregards context which means that it is unable to sufficiently capture complex linguistic features. At the same time, one of the drawbacks of BoW is its high number of dimensions and excessive sparsity. The appearance of word embedding overcomes this shortcoming.

Recently, word embedding based approaches [2,3] have learned from low-dimensional, continuously-valued vector representations using unsupervised methods over the large corpus. State-of-the-art word embedding algorithms include the C and W model [4], the continuous bag-of-words (CBOW) model, the Skip-Gram Word2Vec model [3], and the GloVe (Global Vectors for Word Representation) model [5]. Word embedding techniques have been shown to facilitate a variety of NLP (natural language processing) tasks, including machine translation [6], word analogy [7], POS (part of speech) tagging [8], sequence labeling [4], named entity recognition [9], text classification [10], speech processing [11], and so on. The principle behind these word embedding approaches is the distributional hypothesis that "You shall know a word by the company it keeps" [12]. By leveraging statistical information, such as word co-occurrence frequencies, the method could explicitly encode many linguistic regularities and patterns into vectors. It produces a vector space in which each unique word in the corpus is assigned a corresponding vector in the space, and words with similar contexts in the training corpus are located in close proximity. The feasibility of distribution assumptions has been confirmed in many experiments [13,14].

However, most existing word embedding algorithms only consider statistical information from documents [3,5,7]. The representations learned from these algorithms are not the most effective for emotional analysis tasks. In order to improve the performance of word embedding in emotional analysis tasks, a method that combines the traditional word embedding and provides prior knowledge from external sources is proposed. The knowledge of the polarity and intensity of emotional words can be obtained via public sentiment lexicons, and this sentiment information is not directly obtained in word co-occurrence frequencies, which can greatly enhance the performance of word embedding for emotional analysis. For example, "happy" and "sad" might appear in the same or a similar emotional context but represent different emotions, so it is not enough to learn the emotional information of these two words by counting the word co-occurrence. The method proposed in this paper builds sentiment-aware word embedding by incorporating prior sentiment knowledge into the embedding process.

Our primary contribution is therefore to propose such a solution by making use of the external emotional information, and propose the sentiment-aware word embedding to improve emotional analysis. While there is an abundant literature in the NLP community on word embedding for text representations, much less work has been devoted in comparison to hybrid representation (combining the diverse vector representations into a single representation).The proposed sentiment-aware word embedding is implemented by jointly embedding the word and prior emotional knowledge in the same latent space. The method tests on the NLPCC dataset label Weibo data with seven emotions. First, our method encodes the semantic relationships between words by traditional word embedding. Second, the method incorporates sentiment information of words into emotional embedding. Various combinations of word representations are used in this experiment. It is the hybrid sentiment-aware word embedding that can encode both semantics and sentiments of words. In the experiments, the results show that the two kinds of semantic evidence can complement each other to improve the accuracy of identifying the correct emotion.

The paper is organized as follows. Section 2 briefly introduces the emotion analysis method. Section 3 presents the details of the proposed methodology. Section 4 discusses the experimental arrangement. Section 5 summarizes the contents of the full text and discusses the future direction of development.

2. Related Work

The study of emotional analysis is roughly summed up into three categories: Rule-based analysis, unsupervised classification, and supervised classification.

Rule-based analysis is mainly performed together with the emotion lexicon. In English text analysis, Kamps proposed a distance measurement method to determine the semantic polarity of adjectives based on the synonym graph theory model. In the analysis of Chinese text, Zhu et al. introduced a simple method based on HowNet Chinese lexicon to determine the semantic direction of Chinese words [15]. Pan et al. identified six kinds of emotion expressed by Weibo with the lexicon-based method [16]. However, the lexicon-based method is low in accuracy, and the classifying quality is easily limited by the lexicon [17]. In particular, the lexicon-based method ignores contextual information.

Unsupervised classification analysis does not use tagged documents but relies on a documentation's statistical properties, NLP processes, and existing vocabulary, which has an emotional or polarizing tendency. Turney presented a simple, unsupervised learning algorithm for classifying reviews [18]. Lin and He proposed a novel, probabilistic modeling framework based on latent Dirichlet allocation (LDA), called the joint sentiment/topic model (JST), which detects sentiments and topics simultaneously from the text [19]. They also explored various ways to obtain prior information to improve the accuracy of emotional detection. Yili Wang and Hee Yong Youn [20] proposed a novel feature weighting approach for the sentiment analysis of Twitter data and a fine-grained feature clustering strategy to maximize the accuracy of the analysis.

The analysis based on supervised classification generates an emotion classification model with labeled training data. The effectiveness of supervised technologies depends on the features used in the classification task. The bag-of-words features and their weighting scheme are widely used in natural language processing, which provides a simplified representation of documents through various features. However, these methods have limitations in the task of emotional analysis. Word embedding drives the development of many NLP tasks through the low-dimensional continuous vector representations of words.

In the framework of the word embedding model, the word vector is generated according to the distribution hypothesis [12]. It has been found that learning vectors can clearly encode many linguistic regularities and patterns [21]. However, it is still not enough to rely solely on word-level distribution information collected from the text corpus to learn high-quality representations [22,23]. Auxiliary information has been shown to help learn task-specific word embedding to improve the performance in the tasks [24,25].

In order to improve the representation of word embedding, some research work has been proposed to incorporate various additional resources into the learning framework of word representation. Some knowledge-enhancing word embedding models incorporate lexical knowledge resources into the training process of word embedding models [22,23,26,27]. In the study of Levy and Goldberg [28], the grammatical context of the automatically generated dependency analysis tree was used for word representation training. Meanwhile, some people learn cross-language word embedding with a multilingual parallel corpus [29–31]. Fuji Ren and Jiawen Deng pointed out that background knowledge is composed of keywords and co-occurring words that are extracted from the external corpus and proposed a background knowledge-based multi-stream neural network [32]. Many effective analysis algorithms take advantage of existing knowledge to improve classification performance [33–35]. Wang et al. [33] studied the problem of understanding human sentiments from the large-scale collection of internet images based on both image features and contextual social network information and proved that both visual feature-based and text-based sentiment analysis approaches can learn high-quality models. Tang et al. proposed a learning sentiment-specific word embedding approach dubbed sentiment embedding, which retains the effectiveness of word contexts and exploits the sentiment of text to learn more powerful, continuous word representations [36].

3. Methodology

Constructing an effective features vector to represent text is a basic component in the NLP tasks. In view of the specific emotion classification task, we propose the sentiment-aware word embedding based on the construction of a hybrid word vector method containing emotional information.

This study introduces all of the details about constructing the hybrid feature representations in the following sections (as shown in Figure 1).

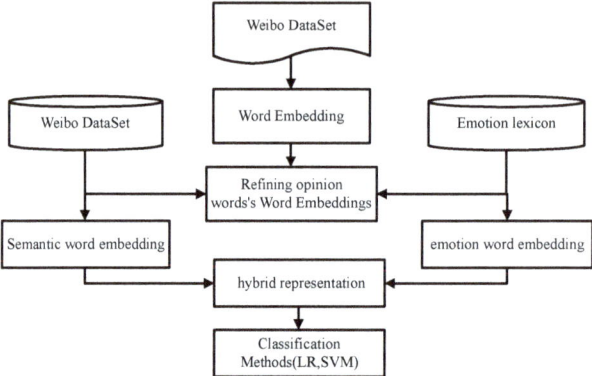

Figure 1. Framework for the proposed method.

The method comprises three main components: (1) the construction of semantic word vectors based on Word2Vec; (2) the construction of emotional word vectors based on the emotional lexicon; and (3) the construction of hybrid sentiment-aware word representations. The third step is based on the results of the previous two steps.

3.1. Constructing a Semantic Word Vector Based on Word2Vec

Word2Vec is an open source tool based on deep learning. It has become more and more popular recently because of its high accuracy in analyzing semantic similarities between two words and the relatively low computational cost. It has two modes: Continuous bag of words (CBOW) and Skip-Gram, which can be applied to quickly learn word embedding from the original text and capture word relationships with the built vector representation model (neural network model). In this paper, the Skip-Gram model was selected for word embedding training. After the text preprocessing step, the word vector representations of all words in the document are learned by the Skip-Gram model, and the vector representation for each word encountered in the input is extracted from the model. Then, the representation of each sentence can be received by averaging over the vectors of all its comprising words. The vector representation of the new document is derived in the same way during the sentiment prediction phase.

3.2. Constructed Emotional Word Vector Based on the Emotional Lexicon

After the preprocessing step, every word in each sentence is converted into a vector. The obtained vectors are then compared based on their cosine similarity degree with the vectors of the emotional words. The emotional word vectors with high similarity scores are selected to be combined. The proposal helps to increase the "semantic power" of traditional semantic space models for emotional analysis by combining different sources of semantic evidence. The hybrid sentiment-aware word embedding is inputted as the experimental model, showing that two kinds of semantic evidence can complement each other, and the mix of them can identify the correct emotion.

The lexicon-based feature extraction method is promising due to the existence of the emotional lexicon. Typically, an emotional lexicon consists of a set of language-specific words, which include information about the emotional category to which it belongs, the polarity intensity, and so on. A fine-grained multi-emotional polarity lexicon can improve the classification accuracy in emotional analysis tasks compared to simple emotional dictionaries [37].

A set of documents and emotional lexicons is given to build the model and generate the emotional word vectors. A corpus, D, consists of a set of texts, $D = \{d_1, d_2 \cdots, d_n\}$, and the vocabulary, $T = \{t_1, t_2 \cdots, t_m\}$, which are unique terms extracted from D. The word representation of the terms t_i are mapped from the Word2Vec model, and then a set of word representations of all words in the vocabulary is derived, i.e., $V = \{v_1, v_2 \cdots, v_m\} \in \mathbb{R}^{m \times d}$, where m is the size of the vocabulary, and d represents the dimensions of the vector space.

The core of this method is to construct emotional word vectors representing the emotional information for each word in the vocabulary, but most of them express some emotions that are not typical opinion words, so the paper introduces the lexicon-based method. In order to construct the emotional word vectors, the method utilizes all emotional words in the lexicon as the emotional vocabulary, $E = [E_1, E_2, ...E_k]$, and gets their word representations, $V_E = [V_{E_1}, V_{E_2}, \cdots, V_{E_k}]$. Due to the scale problem of word embedding the model by training the corpus, there is a low degree of the coverage problem, that is, there are some words that are in the emotional vocabulary but not in the vector spaces learned from Word2Vec. Thus, these words can be ignored and deleted in subsequent processing.

The word embedding model captures the semantic relationships in the corpus text, and the contexts of words with the same emotional polarity are identical. Therefore, the similarity between two items is estimated by calculating the cosine similarity between the vectors represented by V and E, that is, sim $(V, E) = consine\ (V, E)$, defined as

$$\text{sim}(V, E) = \frac{\sum_{i=1}^{d}(V_i * V_{E_i})}{\sqrt{\sum_{i=1}^{d} V_i^2 * \sum_{i=1}^{d} V_{E_i}^2}} \quad (1)$$

where V and E are the vectors of length d.

For each item in T, the similarity to all items in E is calculated and then the similarity results are collected into the matrix $Y \in \mathbb{R}^{m \times k}$, where m is the length of the text glossary, k is the length of the emotional vocabulary, and $Y_{i,j}$ indicates the cosine similarity between the lexical item i and the emotion j. Based on matrix Y, the top n emotional words, $E_w = \{e_1, e_2, \cdots e_n\}$, are selected by setting the threshold. For words in E_w, they are the nearest neighbors to the item in T as determined by distinguishing their sentimental differences.

Similarity in Y means that an item in E and an item in T have the same context; meanwhile, the emotional intensity provided in the dictionary represents the emotional information of the item in E and constructs the emotion vector for all items in T by combining the two kinds of information.

The emotional word vector for each word is calculated as a weighted sum rather than a simple average operation:

$$EmotionVec_i = \frac{1}{n}\sum_{i=1}^{n} weight_i * V_{e_i}. \quad (2)$$

Then, Formula (3) is used to compute the weight of every word; based on this formula, the higher ranked nearest neighbors will receive higher weights:

$$weight_i = Y_{ij} * score_j. \quad (3)$$

The Dalian University of Technology Information Retrieval (DUTIR) is used to represent sentiment lexicon in this study, which is a Chinese ontology resource labeled by the Dalian University of Technology Information Retrieval Laboratory [38]. It contains 27,466 words, and each word is associated with a real-valued score of 1, 3, 5, 7, or 9. The score represents the degree of sentiment, where 9 indicates the maximum strength, and 1 is the minimum strength. Then, we rank the emotional words in E_w by strength scores. The $score_j$ is defined as the reciprocal rank of e_j in the E_w, that is

$$score_j = \frac{1}{rank_j} \tag{4}$$

where $rank_j$ denotes the rank of e_j generated by the intensity-based ranking process.

The weights have to be normalized to sum to one. This can be obtained with the following relation:

$$weight_i = \frac{weight_i}{\sum_{j=1}^{n} weight_j}. \tag{5}$$

In the process of constructing the emotional word vector of words, the shortcomings of the context-based word embedding result in words with opposite sentiment polarity having a fairly high cosine similarity in the vector space. There are great errors generated in the construction of emotional word vectors using word embedding mapping and cosine similarity, so sentimental lexicons are used to correct the word representations mapping from the word embedding space. It is a good way to optimize existing word vectors by using real-valued sentiment intensity scores provided by the emotional lexicons and word vector refinement model. In this way, the words are closer to semantically and emotionally similar words in the dictionary (that is, those with similar intensity scores) and stay away from words that are not emotionally similar but are similar in semantics.

3.3. Constructing Hybrid Sentiment-Aware Word Embedding

In this work, sentiment-aware word embedding is constructed to represent each word by connecting the emotional word vectors based on the lexicon and the semantic word vectors based on the Word2Vec model. Sentiment-aware word embedding can capture the emotional orientation of words, which is the word representation method strictly based on word embedding. In addition, it also makes the most of the contextual and semantic expression ability of the word embedding model.

The sentiment-aware word embedding combines the emotional word vectors with semantic word embedding to simplify combinatorial functions, which indicates that the advantages of the two models can be combined in a single mixed representation. This paper explores different methods of vector combination and experiments with the proposed vector combination method in Section 4. In particular, in order to compare the advantages and limitations of various methods, a comparative study of the two combination methods is conducted.

The first method combines the emotional word vectors with semantic words of a given word directly, which allows two vector representations with different dimension representations:

$$x_{new} = x_e \frown x_s \tag{6}$$

where x_e (x_s) represents the emotional word vector, (semantic word embedding).

Two vectors, x_c and y_e, form by linking the corresponding vectors from the original space x_1 with x_2, y_1, and y_2 as well. Cosine similarity is used to estimate the similarity between two items, and the key factor for cosine similarity is the dot product, i.e.,

$$x_c \cdot y_c = (x_1 \frown x_2) \cdot (y_1 \frown y_2) = \begin{bmatrix} x_1 & x_2 \end{bmatrix} \cdot \begin{bmatrix} y_1 \\ y_2 \end{bmatrix} = x_1 \cdot y_1 + x_2 \cdot y_2. \tag{7}$$

Thus, the cosine similarity in cascade space is determined by the linear combination of the dot products of the vector component. Therefore, the semantic relations and emotional relations between two words are distinguished as features.

The second method is to combine these representational spaces by addition, which requires the two spaces to have the same dimensions, and it can be realized using simple vector addition. The value for each word in the new space is the normalized sum of its two component spaces, i.e.,

$$x_{new} = x_e + x_s. \tag{8}$$

From the dot product result of the vector, the direct superposition of the two vectors which combine the characteristic components of them increases the distinction between different emotional features.

No matter whether vector connection or vector overlay is used, the experimental results show that the advantages of the two models can be combined in a single mixed representation by combining word vector components generated in different ways with simple combinatorial functions.

3.4. Algorithm Implementation

According to data pre-processing, word embedding construction, emotional lexicon processing, and word similarity computation, Algorithm 1 presents the sentiment-aware word embedding approach to emotional analysis. From the algorithm description, it can be seen that with the first iteration, we can get the vocabulary. With the second and third iterations, we can get the vector of each item in the vocabulary and emotional lexicon. Based on the previous step, we can get the hybrid sentiment-aware word embedding through the last iteration.

Algorithm 1 sentiment-aware word embedding for emotion classification

---
Input: a set of unclassified text documents
Output: a set of classified text documents
for *each doc $\in$ corpus* **do**
    for *each term $\in$ doc* **do**
        terms $\leftarrow$ pre-processing all term with NLP methods
    end
    vocabulary $\leftarrow$ terms
end
for *each term $\in$ vocabulary* **do**
    SemanticVector$_{term}$ $\leftarrow$ w2vec$_{term}$
end
for *each emotional_word $\in$ emotional lexicon* **do**
    Vector$_{emotional\_word}$ $\leftarrow$ w2vec$_{emotional\_word}$
end
for *each term $\in$ vocabulary* **do**
    for *each emotional_word $\in$ emotional lexicon* **do**
        sim(term, emotional_word) = consine(term, emotional_word)
        if *sim(term, emotional_word) > Threadhold* **then**
            weight = sim(term, emotional_word) $*$ score(emotional_word)
        end
    end
    EmotionVector$_{term}$ = $\sum_{emotional\_word}$ weigh
    HybridVector1$_{term}$ = SemanticVector$_{term}$ + EmotionVector$_{term}$
    HybridVector2$_{term}$ = SemanticVector$_{term}$ $\frown$ EmotionVector$_{term}$
end

---

4. Experiments

In this section, the proposed hybrid word vector representation method is evaluated with a series of experiments on Weibo text datasets. Based on the above steps, the hybrid word vector is constructed and used as an input to train classification models. The experiments select different approaches to evaluate the proposed hybrid feature vectors. Then, a concrete discussion is given based on different classification results.

The Word2Vec model utilizes the genism software package in Python to create the word vectors, and the Skip-Gram, one method of Word2Vec, was selected for this work. Word2Vec works in a way that is comparable to the machine learning approach of deep neural networks, which allows the representation of semantically similar words in the same vector space with adjacent points.

4.1. Description of Datasets

The data in this paper comes from Weibo (a popular Chinese social networking site). The training corpus for word embedding learning includes 10 million unlabeled blogs of COAE 2014 (The Sixth Chinese Opinion Analysis Evaluation) and 40 thousand blogs labeled using the emotion category in the task of NLPCC2018. The text of Microblog is labeled "none" if it does not convey any emotion. If the text conveys emotion, it is labeled with emotion categories from happiness, trust, anger, sadness, fear, disgust, or surprise. The number of sentences for each emotion category in training data and test data are described in Table 1. We can see from the tables that the distribution of different emotion classes is not balanced.

Table 1. Examples of the emotional vocabulary ontology format.

Emotion Type	Training Data		Test Data	
	Number	Percentage	Number	Percentage
HAPPINESS	5200	13%	3000	15%
TRUST	5200	13%	2200	11%
ANGER	3600	9%	2000	10%
SADNESS	2800	7%	1600	8%
FEAR	2400	6%	1000	5%
DISGUST	4800	12%	2000	10%
SURPRISE	4000	10%	2200	11%
none	12,000	30%	6000	30%

This paper selected DUTIR, which is a Chinese ontology resource collated and labeled by the Dalian University of Technology Information Retrieval Laboratory, as emotional lexicons. There are seven kinds of emotion in the lexical ontology, happiness, trust, anger, sadness, fear, disgust and surprise, a total of 27,466 emotional words, and five emotional intensities, 1, 3, 5, 7, and 9, where 9 indicates the maximum strength and 1 is the least strength. This resource describes a Chinese word or phrase from different perspectives, including lexical categories, emotional categories, emotional intensity, and polarity of words.

In the Chinese Emotion Word Ontology, the general format is shown in Table 2:

Table 2. Examples of the emotional vocabulary ontology format. Part of speech (POS).

Word	POS	Sense	Sense Number	Emotional Subcategory	Strength	Polarity
Happy	adj	2	1	happy	5	1
sad	adj	1	1	sad	5	0

The emotional categories are shown in Table 3.

Table 3. Emotional category.

Emotional Categories	Subcategories	Example Words
HAPPINESS	happy	joy, hilarity
	relieved	relieved, innocent
TRUST	respect	polite, reverent
	praise	handsome, excellent
	believe	trust, reliability
	favorite	admiration, love
	wish	longing for, blessing
ANGER	anger	annoyed, furious
SADNESS	sad	sorrow, heart
	disappointed	regret, despair
	guilty	remorseful, unwilling,
	miss	lovesickness, hangs
FEAR	nervous	flustered, overwhelmed
	dread	timid, scared
	shame	red-faced, self-confident
DISGUST	bored	irritated, upset
	abomination	shameful, hateful
	blame	vanity, chaotic
	jealous	envy
	doubt	suspicious, multi-minded
SURPRISE	surprise	strange, miraculous

4.2. Experimental Results

In this section, we explain the use of the support vector machine (SVM), logistic regression model, decision tree model and gradient boost model classifier to evaluate the effectiveness of the emotional word vector.

The main content is to evaluate the possible advantages of the hybrid word vectors method relative to other word representation methods. The proposed method tries different mixing methods on the emotional word vector and the semantic word vector to compare them according to their classification accuracy on all datasets.

In the emotional classification task, the current widely accepted evaluation indicators are the precision and recall rate. The F-score, a comprehensive metric, was also selected to measure the accuracy of the assessment analysis. The concrete calculation process can be expressed as follows:

$$Precison = \frac{\#system\_correc\,(emotion = i)}{\#system\_proposed\,(emotion = i)} \quad (9)$$

$$Recall = \frac{\#system\_correct\,(emotion = i)}{\#gold\,(emotion = i)} \quad (10)$$

where #$system\_correc(emotion = i)$ represents the number of texts correctly categorized into a class, and #$system\_proposed(emotion = i)$ represents the total number of texts classified into a class. The calculation formula for the F-score is described as follows:

$$F\text{-}measure = \frac{2 \times Precison \times Recall}{Precison + Recall}. \quad (11)$$

Experiments use hybrid vectors as input into different classifiers. Here, the SVM, logistic regression classification model, decision tree model and gradient boost model are selected, and the experimental results can be described as follows.

From Table 4, it can be seen that the classification accuracy and reliability of positive emotions such as happiness and trust are higher than the negative emotions such as fear and disgust. The main reason is that in the training corpus, the text scale of positive emotion is larger than the text scale of negative emotion, for example, the number of texts belonging to happiness and trust accounted for a total of 26% of the training data, while the number of texts belonging to sadness and fear accounted for a total of 11%, which indicates that the effect of classification is related to the distribution of training corpus to a certain extent.

Chinese Weibo text content is short with heavy colloquialism and many novel words on the internet, which limits the process of constructing emotional word vectors based on the emotion dictionary. Despite this, the experiment still enhanced the precision, which further proves the validity of the experimental method. The results of this experiment are related to the quality of emotional dictionaries to a certain extent, and emotional words with strong emotions should be chosen in the course of the experiment.

In order to highlight the effectiveness of the emotional word vector, this study carried out comparative experiments on different word representation methods. Specifically, the performance of the method based on the Skip-Gram and that based on the hybrid word vectors proposed in this paper were compared.

Table 4. Performance evaluation sentiment-aware word embedding. Hybrid method 1 is as shown in Equation (6), hybrid method 2 is as shown in Equation (8), Prec is the abbreviation of Precision which is as shown in Equation (9), Rec is the abbreviation of Recall which is as shown in Equation (10), F1 is the abbreviation of F-measure which is as shown in Equation (11).

Model	Method	Evaluation Indicators	HAPPINESS	TRUST	ANGER	SADNESS	FEAR	DISGUST	SURPRISE
Support vector machine (SVM)	Hybrid method 1	Prec	0.7900	0.7976	0.7901	0.6364	0.6579	0.7273	0.6490
		Rec	0.8315	0.7974	0.7807	0.549	0.4902	0.7141	0.9402
		F1	0.8102	0.7973	0.7793	0.5895	0.5161	0.7093	0.7679
	Hybrid method 2	Prec	0.7733	0.7968	0.7873	0.5714	0.5488	0.6933	0.6393
		Rec	0.7858	0.7965	0.7786	0.4706	0.4794	0.6767	0.9316
		F1	0.7795	0.7964	0.7674	0.5161	0.5118	0.6685	0.7583
Logistic Regression	Hybrid method 1	Prec	0.7600	0.8030	0.7742	0.6314	0.6571	0.7146	0.6377
		Rec	0.7600	0.8029	0.7500	0.5092	0.451	0.7079	0.9301
		F1	0.7600	0.8028	0.7605	0.5652	0.5349	0.7050	0.7500
	Hybrid method 2	Prec	0.7451	0.8030	0.7742	0.5713	0.5908	0.7190	0.6492
		Rec	0.7600	0.8029	0.7500	0.4700	0.5908	0.7192	0.9231
		F1	0.7524	0.8028	0.7605	0.5162	0.5475	0.7104	0.7579
decision_tree	Hybrid method 1	Prec	0.6935	0.6966	0.6157	0.6977	0.5116	0.6543	0.6479
		Rec	0.6900	0.6736	0.6709	0.5128	0.4314	0.6500	0.6900
		F1	0.6917	0.7052	0.6421	0.5911	0.4681	0.6521	0.6682
	Hybrid method 2	Prec	0.6624	0.6522	0.6356	0.6989	0.5080	0.6552	0.6531
		Rec	0.6472	0.6571	0.6410	0.5256	0.538	0.6424	0.6531
		F1	0.6547	0.6546	0.6383	0.6000	0.4512	0.6487	0.6531
gradient_boost	Hybrid method 1	Prec	0.8164	0.7975	0.7900	0.6392	0.6453	0.6344	0.6486
		Rec	0.8034	0.7900	0.8125	0.8632	0.6300	0.8675	0.8205
		F1	0.8087	0.7937	0.8009	0.7345	0.6376	0.7329	0.7245
	Hybrid method 2	Prec	0.7906	0.7900	0.7800	0.6486	0.6420	0.6255	0.6012
		Rec	0.7867	0.7900	0.7786	0.8205	0.6372	0.6496	0.8889
		F1	0.7886	0.7900	0.7792	0.7245	0.6395	0.6373	0.7172

As is shown in the Table 5, the experimental results of hybrid word vectors as classifier input are better than the experimental results of the initial word embedding as classifier input. Among them, hybrid method 1 is slightly better than hybrid method 2. The main reason is that mixing the methods increases the difference between the word vectors of different emotions more significantly. Mixing method 2 also brings about the improvement of precision, but the increase is not significant.

Table 5. Accuracy comparison of emotion classification experiments based on various text representations.

Model	Word Vector	Prec	Rec	F1
svm	Skip-gram	0.6685	0.6811	0.6542
	Hybrid word vector1	0.7211	0.7290	0.7099
	Hybrid word vector2	0.6871	0.7027	0.6865
LR	Skip-gram	0.6612	0.6507	0.6674
	Hybrid word vector1	0.7124	0.7015	0.6969
	Hybrid word vector2	0.6932	0.7166	0.6925
decision_tree	Skip-gram	0.6366	0.6100	0.6230
	Hybrid word vector1	0.6453	0.6170	0.6312
	Hybrid word vector2	0.6379	0.6149	0.6143
gradient_boost	Skip-gram	0.6812	0.7511	0.7145
	Hybrid word vector1	0.7102	0.7981	0.7475
	Hybrid word vector2	0.6968	0.7645	0.7251

4.3. Relation to Other Method

Multi-entity sentiment analysis using entity-level feature extraction and word embeddings approach [39] enhance the word embeddings approach with the deployment of a sentiment lexicon-based technique. The paper proposes associating a given entity with the adjectives, adverbs, and verbs describing it and extracting the associated sentiment to try and infer if the text is positive or negative in relation to the entity or entities. We discuss the major differences between Sweeney's model and our method: (i) Lexicon is used in different ways, Sweeney's model uses the lexicon against the parsed text to identify the polarity of the descriptor words (ii) Sweeney's model uses a Twitter-specific parser to identify the descriptor words that relate to a specific entity for text that contains multiple entities. The descriptor words of the multi-entity tweets are scored using SentiWordNet sentiment lexicon. The overall scoring per entity is printed out as output for the multi-entity tweets. The remaining tweets are classified using a random forest classifier. At the end of the article, the author points out the research aims to highlight how a hybrid word embeddings and lexicon-based approach can be used to tackle the problem of sentiment analysis on multiple entities. But we can see that it isolates dictionary information from word embedding techniques and does not use so-called hybrid word embedding techniques. And our approach is the perfect way to achieve this.

We have reproduced the experiment of the paper and compared it with our method:

As is shown in the Table 6, sentiment-aware word embedding is superior to Sweeney's model in the use of external information, and performs better on emotion classification than Sweeney's model. To sum up, the hybrid word vector that combines semantic information with external emotional information not only provides more word feature information but also involves the emotion labeling of the word in the model prediction. The experiments showed that this method is effective, and it is superior to the original model in terms of accuracy, recall rate, and F-score. It can be concluded that in the emotional analysis task, the quality of the word vector can be improved by incorporating external emotional information.

Table 6. Accuracy comparison of Sweeney's model and sentiment-aware word embedding.

Method	Prec	Rec	F1
SentiWordNet + POS	0.66	0.66	0.66
Word2Vec + SentiWordNet + POS	0.6913	0.6754	0.695
Hybrid word vector1	0.7211	0.729	0.7099
Hybrid word vector2	0.6871	0.7027	0.6865

5. Conclusions

Sentiment-aware word embedding was proposed for emotion classification tasks. The method uses mixed emotional word vectors for emotional analysis, through context-sensitive word embedding provided by Word2Vec combinations of the emotional information provided by the dictionary, which is used as the input of the classifier model. The experiment proved that the use of hybrid word vectors is effective for supervised emotion classification, as it greatly improves the accuracy of emotion classification tasks.

In the future, our work will be aimed at doing some experiments in the following directions to demonstrate the flexibility and effectiveness of the method and to further improve its performance: (a) The method will be applied to the other language corpuses to prove its versatility, and (b) novel ways of combining different word vector components will be explored to increase the differentiation of features between two words.

Author Contributions: F.L. conceived the idea, designed and performed the experiments, and analyzed the results, S.C. drafted the initial manuscript, and S.C., J.S., and X.M. revised the final manuscript. R.S. provided experimental environment and academic guidance.

Funding: This work was supported by the National Natural Science Foundation of China (Grant Nos. 61401519, 61872390), the Natural Science Foundation of Hunan Province (Grant Nos. 2016JJ4119, 2017JJ3415), and the Postdoctoral Science Foundation of China (Grant No. 2016M592450).

Conflicts of Interest: The authors declare no conflict of interest.

References

1. Nikhil, R.; Tikoo, N.; Kurle, S.; Pisupati, H.S.; Prasad, G. A survey on text mining and sentiment analysis for unstructured web data. *J. Emerg. Technol. Innov. Res.* **2015**, *2*, 1292–1296.
2. Huang, E.H.; Socher, R.; Manning, C.D.; Ng, A.Y. Improving word representations via global context and multiple word prototypes. In Proceedings of the Meeting of the Association for Computational Linguistics: Long Papers, Jeju Island, Korea, 8–14 July 2012; pp. 873–882.
3. Mikolov, T.; Chen, K.; Corrado, G.; Dean, J. Efficient Estimation of Word Representations in Vector Space. *arXiv* **2013**, arXiv:1301.3781.
4. Collobert, R.; Weston, J.; Karlen, M.; Kavukcuoglu, K.; Kuksa, P. Natural Language Processing (Almost) from Scratch. *J. Mach. Learn. Res.* **2011**, *12*, 2493–2537.
5. Pennington, J.; Socher, R.; Manning, C. Glove: Global vectors for word representation. In Proceedings of the 2014 Conference on Empirical Methods in Natural Language Processing (EMNLP), Doha, Qatar, 25–29 October 2014; pp. 1532–1543.
6. Devlin, J.; Zbib, R.; Huang, Z.; Lamar, T.; Schwartz, R.; Makhoul, J. Fast and Robust Neural Network Joint Models for Statistical Machine Translation. In Proceedings of the 52nd Annual Meeting of the Association for Computational Linguistics, Baltimore, MD, USA, 23–25 June 2014; pp. 1370–1380.
7. Mikolov, T.; Sutskever, I.; Chen, K.; Corrado, G.; Dean, J. Distributed Representations of Words and Phrases and their Compositionality. In Proceedings of the Advances in Neural Information Processing Systems, Lake Tahoe, NV, USA, 5–10 December 2013; pp. 3111–3119.
8. Lin, C.-C.; Ammar, W.; Dyer, C.; Levin, L. Unsupervised pos induction with word embeddings. *arXiv* **2015**, arXiv:1503.06760.
9. Turian, J.; Ratinov, L.; Bengio, Y. Word representations: A simple and general method for semi-supervised learning. In Proceedings of the Meeting of the Association for Computational Linguistics, ACL 2010, Uppsala, Sweden, 11–16 July 2010; pp. 384–394.
10. Joulin, A.; Grave, E.; Bojanowski, P.; Mikolov, T. Bag of Tricks for Efficient Text Classification. *arXiv* **2016**, arXiv:1607.01759.
11. Mesnil, G.; Dauphin, Y.; Yao, K.; Bengio, Y.; Deng, L.; Hakkanitur, D.; He, X.; Heck, L.; Tur, G.; Yu, D. Using Recurrent Neural Networks for Slot Filling in Spoken Language Understanding. *IEEE/ACM Trans. Audio Speech Lang. Process.* **2015**, *23*, 530–539. [CrossRef]
12. Harris, Z.S. *Distributional Structure*; Springer: Dordrecht, The Netherlands, 1981; pp. 146–162.
13. Charles, W.G. Contextual correlates of meaning. *Appl. Psycholinguist.* **2000**, *21*, 505–524. [CrossRef]

14. Rubenstein, H.; Goodenough, J.B. Contextual correlates of synonymy. *Commun. ACM* **1965**, *8*, 627–633. [CrossRef]
15. Zhu, Y.; Min, J.; Zhou, Y. Semantic orientation computing based on HowNet. *J. Chin. Inf. Process.* **2006**, *20*, 14–20.
16. Pan, M.H.; Niu, Y. Emotion Recognition of Micro-blogs Based on a Hybrid Lexicon. *Comput. Technol. Dev.* **2014**, *9*, 6.
17. Cambria, E.; Schuller, B.; Xia, Y.; Havasi, C. New Avenues in Opinion Mining and Sentiment Analysis. *IEEE Intell. Syst.* **2013**, *28*, 15–21. [CrossRef]
18. Turney, P.D. Thumbs up or thumbs down?: Semantic orientation applied to unsupervised classification of reviews. In Proceedings of the 40th annual meeting on association for computational linguistics, Philadelphia, PA, USA, 7–12 July 2002; pp. 417–424.
19. Lin, C.; He, Y. Joint sentiment/topic model for sentiment analysis. In Proceedings of the 18th ACM Conference on Information and Knowledge Management, Hong Kong, China, 2–6 November 2009; pp. 375–384.
20. Wang, Y.; Youn, H. Feature Weighting Based on Inter-Category and Intra-Category Strength for Twitter Sentiment Analysis. *Appl. Sci.* **2019**, *9*, 92. [CrossRef]
21. Mikolov, T.; Yih, W.-t.; Zweig, G. Linguistic regularities in continuous space word representations. In Proceedings of the 2013 Conference of the North American Chapter of the Association for Computational Linguistics: Human Language Technologies, Atlanta, GA, USA, 9–14 June 2013; pp. 746–751.
22. Liu, Q.; Jiang, H.; Wei, S.; Ling, Z.-H.; Hu, Y. Learning semantic word embeddings based on ordinal knowledge constraints. In Proceedings of the 53rd Annual Meeting of the Association for Computational Linguistics and the 7th International Joint Conference on Natural Language Processing, Beijing, China, 26–31 July 2015; pp. 1501–1511.
23. Faruqui, M.; Dodge, J.; Jauhar, S.K.; Dyer, C.; Hovy, E.; Smith, N.A. Retrofitting Word Vectors to Semantic Lexicons. *arXiv* **2014**, arXiv:1411.4166.
24. Liu, Y.; Liu, Z.; Chua, T.S.; Sun, M. Topical word embeddings. In Proceedings of the Twenty-Ninth AAAI Conference on Artificial Intelligence, Austin, TX, USA, 25–30 January 2015; pp. 2418–2424.
25. Zhou, C.; Sun, C.; Liu, Z.; Lau, F.C.M. Category Enhanced Word Embedding. *arXiv* **2015**, arXiv:1511.08629.
26. Yu, M.; Dredze, M. Improving Lexical Embeddings with Semantic Knowledge. In Proceedings of the 52nd Annual Meeting of the Association for Computational Linguistics, Baltimore, MD, USA, 23–25 June 2014; pp. 545–550.
27. Xu, C.; Bai, Y.; Bian, J.; Gao, B.; Wang, G.; Liu, X.; Liu, T.Y. RC-NET: A General Framework for Incorporating Knowledge into Word Representations. In Proceedings of the ACM International Conference on Conference on Information and Knowledge Management, Shanghai, China, 3–7 November 2014; pp. 1219–1228.
28. Levy, O.; Goldberg, Y. Dependency-Based Word Embeddings. In Proceedings of the 52nd Annual Meeting of the Association for Computational Linguistics, Baltimore, MD, USA, 23–25 June 2014; pp. 302–308.
29. Lu, A.; Wang, W.; Bansal, M.; Gimpel, K.; Livescu, K. Deep Multilingual Correlation for Improved Word Embeddings. In Proceedings of the Conference of the North American Chapter of the Association for Computational Linguistics: Human Language Technologies, Denver, CO, USA, 31 May–5 June 2015; pp. 250–256.
30. Hermann, K.M.; Blunsom, P. Multilingual Models for Compositional Distributed Semantics. *arXiv* **2014**, arXiv:1404.4641.
31. Zhang, J.; Liu, S.; Li, M.; Zhou, M.; Zong, C. Bilingually-constrained phrase embeddings for machine translation. In Proceedings of the 52nd Annual Meeting of the Association for Computational Linguistics, Baltimore, MD, USA, 23–25 June 2014; pp. 111–121.
32. Ren, F.; Deng, J. Background Knowledge Based Multi-Stream Neural Network for Text Classification. *Appl. Sci.* **2018**, *8*, 2472. [CrossRef]
33. Wang, Y.; Wang, S.; Tang, J.; Liu, H.; Li, B. Unsupervised sentiment analysis for social media images. In Proceedings of the IEEE International Conference on Data Mining Workshop, Washington, DC, USA, 14–17 November 2015; pp. 1584–1591.
34. Hogenboom, A.; Bal, D.; Frasincar, F.; Bal, M.; Jong, F.D.; Kaymak, U. Exploiting emoticons in sentiment analysis. In Proceedings of the 28th Annual ACM Symposium on Applied Computing, Coimbra, Portugal, 18–22 March 2013; pp. 703–710.

35. Hu, X.; Tang, J.; Gao, H.; Liu, H. Unsupervised sentiment analysis with emotional signals. In Proceedings of the 22nd International Conference on World Wide Web, Rio de Janeiro, Brazil, 13–17 May 2013; pp. 607–618.
36. Tang, D.; Wei, F.; Qin, B.; Yang, N.; Liu, T.; Zhou, M. Sentiment Embeddings with Applications to Sentiment Analysis. *IEEE Trans. Knowl. Data Eng.* **2016**, *28*, 496–509. [CrossRef]
37. Carrillo-De-Albornoz, J.; Plaza, L. An emotion-based model of negation, intensifiers, and modality for polarity and intensity classification. *J. Assoc. Inf. Sci. Technol.* **2013**, *64*, 1618–1633. [CrossRef]
38. Chen, J. The Construction and Application of Chinese Emotion Word Ontology. Master's Thesis, Dailian University of Technology, Dalian, China, 2008.
39. Sweeney, C.; Padmanabhan, D. Multi-entity sentiment analysis using entity-level feature extraction and word embeddings approach. In Proceedings of the Recent Advances in Natural Language Processing, Varna, Bulgaria, 4–6 September 2017.

© 2019 by the authors. Licensee MDPI, Basel, Switzerland. This article is an open access article distributed under the terms and conditions of the Creative Commons Attribution (CC BY) license (http://creativecommons.org/licenses/by/4.0/).

Article

A Deep Learning-Based Approach for Multi-Label Emotion Classification in Tweets

Mohammed Jabreel [1,2,*] and Antonio Moreno [1]

1 ITAKA Research Group, Universitat Rovira i Virgili, 43007 Tarragona, Spain; antonio.moreno@urv.cat
2 Department of Computer Science, Hodeidah University, 1821 Hodeidah, Yemen
* Correspondence: mhjabreel@gmail.com

Received: 15 February 2019; Accepted: 12 March 2019; Published: 17 March 2019

Abstract: Currently, people use online social media such as Twitter or Facebook to share their emotions and thoughts. Detecting and analyzing the emotions expressed in social media content benefits many applications in commerce, public health, social welfare, etc. Most previous work on sentiment and emotion analysis has only focused on single-label classification and ignored the co-existence of multiple emotion labels in one instance. This paper describes the development of a novel deep learning-based system that addresses the multiple emotion classification problem in Twitter. We propose a novel method to transform it to a binary classification problem and exploit a deep learning approach to solve the transformed problem. Our system outperforms the state-of-the-art systems, achieving an accuracy score of 0.59 on the challenging SemEval2018 Task 1:E-c multi-label emotion classification problem.

Keywords: opinion mining; sentiment analysis; emotion classification; deep learning; Twitter

1. Introduction

Emotions are the key to people's feelings and thoughts. Online social media, such as Twitter and Facebook, have changed the language of communication. Currently, people can communicate facts, opinions, emotions, and emotion intensities on different kinds of topics in short texts. Analyzing the emotions expressed in social media content has attracted researchers in the natural language processing research field. It has a wide range of applications in commerce, public health, social welfare, etc. For instance, it can be used in public health [1,2], public opinion detection about political tendencies [3,4], brand management [5], and stock market monitoring [6]. Emotion analysis is the task of determining the attitude towards a target or topic. The attitude can be the polarity (positive or negative) or an emotional state such as joy, anger, or sadness [7].

Recently, the multi-label classification problem has attracted considerable interest due to its applicability to a wide range of domains, including text classification, scene and video classification, and bioinformatics [8]. Unlike the traditional single-label classification problem (i.e., multi-class or binary), where an instance is associated with only one label from a finite set of labels, in the multi-label classification problem, an instance is associated with a subset of labels.

Most previous work on sentiment and emotion analysis has only focused on single-label classification. Hence, in this article, we focus on the multi-label emotion classification task, which aims to develop an automatic system to determine the existence in a text of none, one, or more out of eleven emotions: the eight Plutchik [9] categories (joy, sadness, anger, fear, trust, disgust, surprise, and anticipation) that are shown in Figure 1, plus optimism, pessimism, and love.

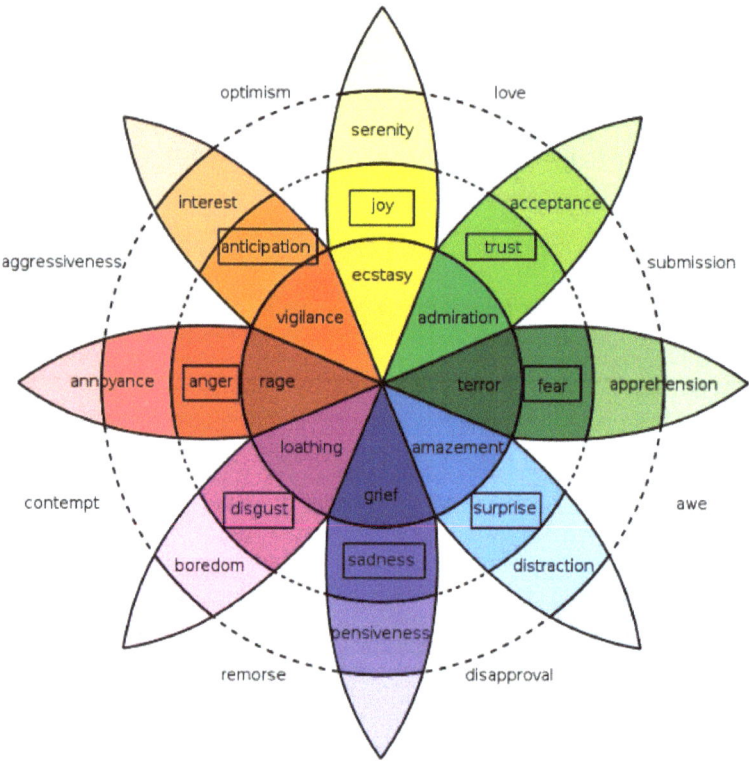

Figure 1. The set of the eight basic emotions proposed by Plutchik [9].

One of the most common approaches to addressing the problem of multi-label classification is the *problem transformation*. With this approach, a multi-label problem is transformed into one or more single-label (i.e., binary or multi-class) problems. Specifically, single-label classifiers are learned and employed; after that, the classifiers' predictions are transformed into multi-label predictions.

Different transformation methods have been proposed in the multi-label literature. The most common method is called *binary relevance* [10,11]. The idea of the binary relevance method is simple and intuitive. A multi-label problem is transformed into multiple binary problems, one problem for each label. Then, an independent binary classifier is trained to predict the relevance of one of the labels. Although binary relevance is popular in the literature, due to its simplicity, it suffers from directly modeling correlations that may exist between labels. However, it is highly resistant to overfitting label combinations, since it does not expect examples to be associated with previously-observed combinations of labels.

In this article, we propose a novel transformation method, called xy-pair-set, for the multi-label classification problem. Unlike binary relevance methods, our method transforms the problem into only one binary classification problem as described in Section 3. Additionally, we exploit the successes of deep learning models, especially the word2vecmethods' family [12] and the recurrent neural networks [13,14] and attention models [15,16], to develop a system that solves the transformed binary classification problem. The critical component of our system is the embedding module, which uses three embedding models and an attention function to model the relationship between the input and the label.

To summarize, the contribution of this work is four-fold.

- We propose a novel transformation mechanism for the multi-label classification problem.
- We propose a novel, attentive deep learning system, which we call Binary Neural Network (BNet), which works on the new transformation method. Our system is a data-driven, end-to-end neural-based model, and it does not rely on external resources such as parts of speech taggers and sentiment or emotion lexicons.
- We evaluate the proposed system on the challenging multi-label emotion classification dataset of SemEval-2018 Task1: Affect in Tweets.
- The experimental results show that our system outperforms the state-of-the-art systems.

The rest of the article is structured as follows. In Section 2, we overview the related work on multi-label problem transformation methods and Twitter sentiment and emotion analysis. In Section 3, we explain in detail the methodology. In Section 4, we report the experimental results. In Section 5, the conclusions and the future work are presented.

2. Related Works

In this section, we overview the most popular research studies related to this work. In Section 2.1, we summarize the most common multi-label problem transformation methods. Section 2.2 gives an overview of the state-of-the-art works on the problem of multi-label emotion classification on Twitter.

2.1. Problem Transformation Methods

Let $X = \{x_1, x_2, \ldots x_n\}$ be the set of all instances and $Y = \{y_1, y_2, \ldots, y_m\}$ be the set of all labels. We can define the set of data:

$$D = \{(x_i, \hat{Y}_i) | x_i \in X \text{ and } \hat{Y}_i \subseteq Y \text{ is the set of labels associated with } x_i\} \tag{1}$$

In this expression, D is called a supervised multi-label dataset.

The task of multi-label classification is challenging because the number of label sets grows exponentially as the number of class labels increases. One common strategy to address this issue is to transform the problem into a traditional classification problem. The idea is to simplify the learning process by exploiting label correlations. Based on the order of the correlations, we can group the existing transformation methods into three approaches [17,18], namely first-order approaches, second-order approaches, and high-order approaches.

First-order approaches decompose the problem into some independent binary classification problems. In this case, one binary classifier is learned for each possible class, ignoring the co-existence of other labels. Thus, the number of independent binary classifiers needed is equal to the number of labels. For each multi-label training example $(x_i, \hat{Y}_i) \in D$, $y_k \in Y$, we construct a binary classification training set, D_k as the following: x_i will be regarded as one positive example if $y_k \in \hat{Y}_i$ and one negative example otherwise. In the first case, we will get a training example in the form $(x_i, 1) \in D_k$, which will be $(x_i, 0) \in D_k$ in the second case. Thus, for all labels $\{y_1, y_2, \ldots, y_m\} \in Y$, m training sets $\{D_1, D_2, \ldots, D_m\}$ are constructed. Based on that, for each training set D_k, one binary classifier can be learned with popular learning techniques such as AdaBoost [19], k-nearest neighbor [20], decision trees, random forests [21,22], etc. The main advantage of first-order approaches is their conceptual simplicity and high efficiency. However, these approaches can be less effective due to their ignorance of label correlations.

Second-order approaches try to address the lack of modeling label correlations by exploiting pairwise relationships between the labels. One way to consider pairwise relationships is to train one binary classifier for each *pair* of labels [23]. Although second-order approaches perform well in several domains, they are more complicated than the first-order approaches in terms of the number of classifiers. Their complexity is quadratic, as the number of classifiers needed is $\binom{m}{2}$. Moreover, in real-world applications, label correlations could be more complex and go beyond second-order.

High-order approaches tackle the multi-label learning problem by exploring high-order relationships among the labels. This can be fulfilled by assuming linear combinations [24], a nonlinear mapping [25,26], or a shared subspace over the whole label space [27]. Although high-order approaches have stronger correlation-modeling capabilities than their first-order and second-order counterparts, these approaches are computationally demanding and less scalable.

Our transformation mechanism, shown in Section 3.1, is a simple as the first-order approaches and can model, implicitly, high-order relationships among the labels if some requirements, detailed in Section 3.1, are fulfilled. It requires only one binary classifier, and the number of training examples grows polynomially in terms of the number of instances and the number of labels. If the number of training examples in the multi-label training dataset is n and the number of the labels is m, then the number of the training examples in the transformed binary training set is $n \times m$.

2.2. Emotion Classification in Tweets

Various machine learning approaches have been proposed for traditional emotion classification and multi-label emotion classification. Most of the existing systems solve the problem as a text classification problem. Supervised classifiers are trained on a set of annotated corpora using a different set of hand-engineered features. The success of such models is based on two main factors: a large amount of labeled data and the intelligent design of a set of features that can distinguish between the samples. With this approach, most studies have focused on engineering a set of efficient features to obtain a good classification performance [28–30]. The idea is to find a set of informative features to reflect the sentiments or the emotions expressed in the text. Bag-of-Words (BoW) and its variation, n-grams, is the representation method used in most text classification problems and emotion analysis. Different studies have combined the BoW features with other features such as the parts of speech tags, the sentiment and the emotion information extracted from lexicons, statistical information, and word shapes to enrich the text representation.

Although BoW is a popular method in most text classification systems, it has some drawbacks. Firstly, it ignores the word order. That means that two documents may have the same or a very close representation as far as they have the same words, even though they carry a different meaning. The n-gram method resolves this disadvantage of BoW by considering the word order in a context of length n. However, it suffers from sparsity and high dimensionality. Secondly, BoW is scarcely able to model the semantics of words. For example, the words *beautiful*, *wonderful* and *view* have an equal distance in BoW, where the word *beautiful* is closer to the word *wonderful* than the word *view* in the semantic space.

Sentiment and emotion lexicons play an essential role in developing efficient sentiment and emotion analysis systems. However, it is difficult to create such lexicons. Moreover, finding the best combination of lexicons in addition to the best set of statistical features is a time-consuming task.

Recently, deep learning models have been utilized to develop end-to-end systems in many tasks including speech recognition, text classification, and image classification. It has been shown that such systems automatically extract high-level features from raw data [31,32].

Baziotis et al. [33], the winner of the multi-label emotion classification task of SemEval-2018 Task1: Affect in Tweets, developed a bidirectional Long Short-Term Memory (LSTM) with a deep attention mechanism. They trained a word2vec model with 800,000 words derived from a dataset of 550 million tweets. The second place winner of the SemEval leaderboard trained a word-level bidirectional LSTM with attention, and it also included non-deep learning features in its ensemble [34]. Ji Ho Park et al. [35] trained two models to solve this problem: regularized linear regression and logistic regression classifier chain [11]. They tried to exploit labels' correlation to perform multi-label classification. With the first model, the authors formulated the multi-label classification problem as a linear regression with label distance as the regularization term. In their work, the logistic regression classifier chain method was used to capture the correlation of emotion labels. The idea is to treat the multi-label problem as a sequence of binary classification problems by taking the prediction of the previous classifier as an extra input to the next classifier.

In this work, we exploited the deep learning-based approach to develop a system that can extract a high-level representation of the tweets and model an implicit high-order relationship among the labels. We used the proposed system alongside the proposed transformation method to train a function that can solve the problem of multi-label emotion classification in tweets. The next section explains the details of our proposed system.

3. Methodology

This section shows the methodology of this work. First, we explain in Section 3.1 the proposed transformation method, xy-pair-set. Afterwards, we describe the proposed system in Section 3.2.

3.1. xy-Pair-Set: Problem Transformation

The proposed transformation method xy-pair-set transforms a multi-label classification dataset D into a supervised binary dataset $\hat{D}$ as follows:

$$\forall x_i \in X, y \in Y \text{ and } (x_i, \hat{Y}_i) \in D, \exists!((x_i, y), \phi) \in \hat{D}$$

$$\text{where } \phi = \begin{cases} 1 & \text{if } y \in \hat{Y}_i \\ 0 & \text{otherwise} \end{cases} \tag{2}$$

Algorithm 1 explains the implementation of the proposed transformation method. It takes as inputs a multi-label dataset D (Equation (1)) and a set of labels Y, and it returns a transformed binary dataset. We show next an illustrative example.

Algorithm 1: xy-pair-set algorithm.

Input: Input: a multi-label classification dataset D and a set of labels Y
Output: Output: a binary classification dataset $\hat{D}$

1   $\hat{D} = \{\}$;
2   **foreach** $(x_i, \hat{Y}_i) \in D$ **do**
3   **foreach** $y \in Y$ **do**
4          $\hat{x}_i = (x_i, y)$;   ▷ a tuple of x_i and y.
5   **if** $y \in \hat{Y}_i$ **then**
6              $\hat{D} = \hat{D} \cup (\hat{x}_i, 1)$;
7   **else**
8              $\hat{D} = \hat{D} \cup (\hat{x}_i, 0)$;
9   **end**
10   **end**
11 **end**
12 **return** $\hat{D}$

Let $X = \{x1, x2\}$, $Y = \{a, b, c\}$ and $D = \{(x1, \{a, c\}), (x2, \{b\})\}$. The output of the binary relevance transformation method is a set of three independent binary datasets, one for each label. That is, $D_a = \{(x1, 1), (x2, 0)\}$, $D_b = \{(x1, 0), (x2, 1)\}$, and $D_c = \{(x1, 1), (x2, 0)\}$. In contrast, the output of our transformation method is a single binary dataset $\hat{D} = \{((x1, a), 1), ((x1, b), 0), ((x1, c), 1), ((x2, a), 0), ((x2, b), 1), ((x2, c), 0)\}$.

The task in this case, unlike the traditional supervised binary classification algorithms, is to develop a learning algorithm to learn a function $g : X \times Y \to \{0, 1\}$. The success of such an algorithm is based on three requirements: (1) an encoding method to represent an instance $x \in X$ as a high-dimensional vector V_x, (2) a method to encode a label $y \in Y$ as a vector V_y, and (3) a method to represent the relation between the instance x and the label y. These three conditions make g able to capture the relationships inputs-to-labels and labels-to-labels. In this work, we take advantage of the successes of deep learning models to fulfill the three requirements listed above. We empirically show the success of our system with respect to these conditions as reported in Sections 4.6 and 4.7.

3.2. BNet: System Description

This subsection explains the proposed system to solve the transformed binary problem mentioned above. Figure 2 shows the graphical depiction of the system's architecture. It is composed of three parts: the embedding module, the encoding module, and the classification module. We explain in detail each of them below.

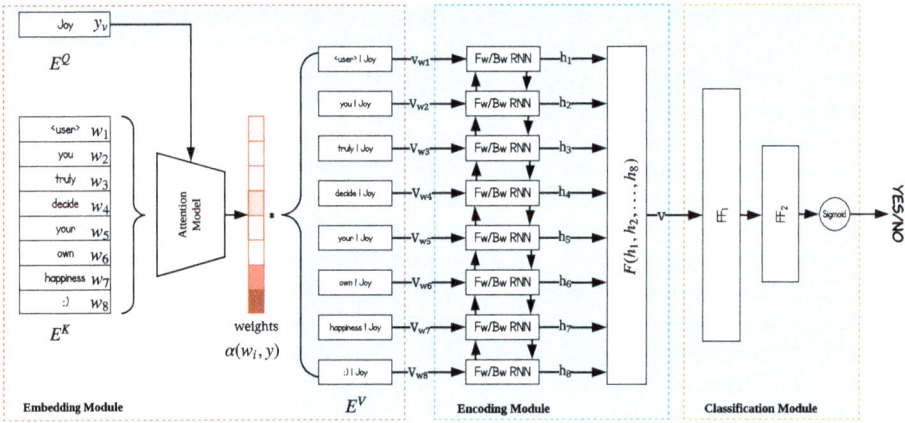

Figure 2. An illustration of the proposed system. Shaded parts are trainable. Fw and Bw refer to the Forward and Backward cells respectively, and FF means Feed-Forward layer.

3.2.1. Embedding Module

Let (W, y) be the pair of inputs to our system, where $W = \{w_1, w_2, \ldots, w_l\}$ is the set of the words in a tweet and y is the label corresponding to an emotion. The goal of the embedding module is to represent each word w_i by a vector v_{w_i} and the label by a vector v_y.

Our embedding module can be seen as a function that maps a query and a set of key-value pairs to an output, where the query, keys, values, and output are all vectors. The output is computed as a weighted sum of the values, where the weight assigned to each value is computed by a compatibility function of the query with the corresponding key. The query is the trainable label embedding, $E^Q(y)$; the keys are the pretrained words embeddings, $E^K(w_i) \; \forall w_i \in W$; and the values are the trainable words embeddings, $E^V(w_i) \; \forall w_i \in W$.

As shown in Figure 2, we used the output of E^Q and E^K as inputs to the attention model to find the alignments, i.e., the weights α, between the label y and the words W of the input tweet. This step models the relation between the input and the label. As soon as the weights were obtained, we then multiplied each word's vector that came from the embedding E^V by its corresponding weight. Given that, the final representation of a word $w_i \in W$ is given as the following:

$$v_{w_i} = E^V(w_i) \cdot \alpha(w_i, y) \qquad (3)$$

The function α is an attention-based model, which finds the strength of the relationship between the word w_i and the label y based on their semantic similarity. That is, $\alpha(w_i, y)$ is a value based on the distance $\Delta(w_i, y)$ between w_i and y as:

$$\alpha(w_i, y) = \frac{e^{\Delta(w_i,y)}}{\sum_{w_j \in W} e^{\Delta(w_j,y)}} \quad (4)$$

Here, $\Delta(w_i, y)$ is a scalar score that represents the similarity between the word w_i and the label y:

$$\Delta(w_i, y) = E^k(w_i) \cdot v_y^T \quad (5)$$

$$v_y = E^Q(y) \quad (6)$$

It is worth noting that $\alpha(w_i, y) \in [0, 1]$ and:

$$\sum_{w_i \in W} \alpha(w_i, y) = 1 \quad (7)$$

3.2.2. Encoding Module

The goal of the encoding module is to map the sequence of word representations $\{v_{w_1}, v_{w_2}, \ldots, v_{w_l}\}$ that is obtained from the embedding module to a single real-valued dense vector. In this work, we used a Recurrent Neural Network (RNN) to design our encoder. RNN reads the input sequence of vectors in a forward direction (left-to-right) starting from the first symbol v_{w_1} to the last one v_{w_l}. Thus, it processes sequences in temporal order, ignoring the future context. However, for many tasks on sequences, it is beneficial to have access to future, as well as to past information. For example, in text processing, decisions are usually made after the whole sentence is known. The Bidirectional Recurrent Neural Network (BiRNN) variant [13] proposed a solution for making predictions based on both past and future information.

A BiRNN consists of forward $\overrightarrow{\phi}$ and backward $\overleftarrow{\phi}$ RNNs. The first one reads the input sequence in a forward direction and produces a sequence of forward hidden states $(\overrightarrow{h_1}, \ldots, \overrightarrow{h_l})$, whereas the former reads the sequence in the reverse order $(v_{w_l}, \ldots, v_{w_1})$, resulting in a sequence of backward hidden states $(\overleftarrow{h_l}, \ldots, \overleftarrow{h_1})$.

We obtained a representation for each word v_{w_t} by concatenating the corresponding forward hidden state $\overrightarrow{h_t}$ and the backward one $\overleftarrow{h_t}$. The following equations illustrate the main ideas:

$$\overrightarrow{h_t} = \overrightarrow{\phi}(v_{w_t}, \overrightarrow{h_{t-1}}) \quad (8)$$

$$\overleftarrow{h_t} = \overleftarrow{\phi}(v_{w_t}, \overleftarrow{h_{t+1}}) \quad (9)$$

$$h_t = [\overrightarrow{h_t}; \overleftarrow{h_t}] \quad (10)$$

The final input representation of the sequence is:

$$c = F(\{h_1, h_2, \ldots, h_l\}) \quad (11)$$

We simply chose F to be the last hidden state (i.e., $F(\{h_1, h_2, \ldots, h_n\}) = h_n$).

In this work, we used two Gated Recurrent Units (GRUs) [14], one as $\overrightarrow{\phi}$ and the other as $\overleftarrow{\phi}$. This kind of RNN was designed to have more persistent memory, making them very useful to capture long-term dependencies between the elements of a sequence. Figure 3 shows a graphical depiction of a gated recurrent unit.

A GRU has *reset* (r_t) and *update* (z_t) gates. The former can completely reduce the past hidden state h_{t-1} if it finds that it is irrelevant to the computation of the new state, whereas the later is responsible for determining how much of h_{t-1} should be carried forward to the next state h_t.

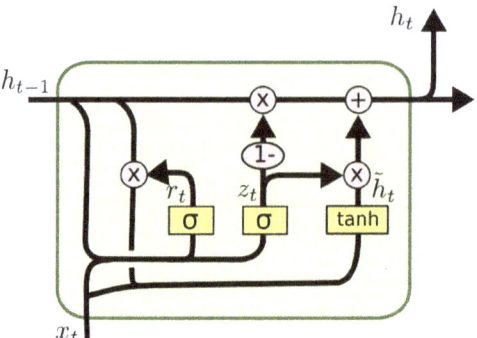

Figure 3. Gated Recurrent Unit (GRU).

The output h_t of a GRU depends on the input x_t and the previous state h_{t-1}, and it is computed as follows:

$$r_t = \sigma\left(W_r \cdot [h_{t-1}; x_t] + b_r\right) \tag{12}$$

$$z_t = \sigma\left(W_z \cdot [h_{t-1}; x_t] + b_z\right) \tag{13}$$

$$\tilde{h}_t = \tanh\left(W_h \cdot [(r_t \odot h_{t-1}); x_t] + b_h\right) \tag{14}$$

$$h_t = (1 - z_t) \odot h_{t-1} + z_t \odot \tilde{h}_t \tag{15}$$

In these expressions, r_t and z_t denote the *reset* and *update* gates, $\tilde{h}_t$ is the candidate output state, and h_t is the actual output state at time t. The symbol $\odot$ stands for element-wise multiplication; σ is a sigmoid function; and; stands for the vector-concatenation operation. $W_r, W_z, W_h \in \mathbb{R}^{d_h \times (d + d_h)}$ and $b_r, b_z, b_h \in \mathbb{R}^{d_h}$ are the parameters of the *reset* and *update* gates, where d_h is the dimension of the hidden state and d is the dimension of the input vector.

3.2.3. Classification Module

Our classifier was composed of two feed-forward layers with the *ReLU* activation function followed by a Sigmoid unit.

4. Experiments and Results

In this section, we first describe the experimental details, and then, we describe the dataset and the pre-processing we used. Afterwards, we introduce the state-of-the-art systems we compared our system with, and finally, we report the empirical validation proving the effectiveness of our system.

4.1. Experimental Details

Table 1 shows the hyperparameters of our system, which was trained using Adam [36], with a learning rate of 0.0001, $\beta_1 = 0.5$, and a mini-batch size of 32 to minimize the binary cross-entropy loss function:

$$\mathcal{L}(\theta, \hat{D}) = -\mathbb{E}_{((x_i, y_i), \phi_i) \sim \hat{D}} \left[\phi_i \cdot g(x_i, y_i) + (1 - \phi_i) \cdot (1 - g(x_i, y_i))\right] \tag{16}$$

where, $g(x_i, y_i)$ is the predicted value, ϕ_i is the real value, and θ is the model's parameters.

The hyperparameters of our system were obtained by applying Bayesian optimization [37]. We used the development set as a validation set to fine-tune those parameters.

Table 1. Hyperparameters of our system.

Parameter	Value
Embedding Module	E^Q: Dimensions: 11 × 310 Initialization: Uniform (−0.02, 0.02) Trainable: Yes E^K: Dimensions: 13,249 × 310 Initialization: Pretrained model (we used the pretrained embeddings provided in [33]). Trainable: No E^V: Dimensions: 13,249 × 310 Initialization: Uniform (−0.02, 0.02) Trainable: Yes
Encoding Module	RNN Cell: GRU Hidden size: 200 Layers: 2 Encoding: last hidden state RNN dropout: 0.3
Classification Module	FF1: 1024 units FF2: 512 units Sigmoid: 1 unit Activation: ReLU Dropout: 0.3

4.2. Dataset

In our experiments, we used the multi-label emotion classification dataset of SemEval-2018 Task1: Affect in Tweets [30]. It contains 10,983 samples divided into three splits: training set (6838 samples), validation set (886 samples), and testing set (3259 samples). For more details about the dataset, we refer the reader to [38]. We trained our system on the training set and used the validation set to fine-tune the parameters of the proposed system. We pre-processed each tweet in the dataset as follows:

- Tokenization: We used an extensive list of regular expressions to recognize the following meta information included in tweets: Twitter markup, emoticons, emojis, dates, times, currencies, acronyms, hashtags, user mentions, URLs, and words with emphasis.
- As soon as the tokenization was done, we lowercased words and normalized the recognized tokens. For example, URLs were replaced by the token "<URL>", and user mentions were replaced by the token "<USER>". This step helped to reduce the size of the vocabulary without losing information.

4.3. Comparison with Other Systems

We compared the proposed system with the state-of-the-art systems used in the task of multi-label emotion classification, including:

- SVM-unigrams: a baseline support vector machine system trained using just word unigrams as features [30].
- NTUA-SLP: the system submitted by the winner team of the SemEval-2018 Task1:E-cchallenge [33].
- TCS: the system submitted by the second place winner [34].
- PlusEmo2Vec: the system submitted by the third place winner [35].
- Transformer: a deep learning system based on large pre-trained language models developed by the NVIDIA AI lab [39].

4.4. Evaluation Metrics

We used multi-label accuracy (or Jaccard index), the official competition metric used by the organizers of SemEval-2018 Task 1: Affect in Tweets, for the E-c sub task, which can be defined as the size of the intersection of the predicted and gold label sets divided by the size of their union.

$$Jaccard = \frac{1}{|T|} \sum_{t \in T} \frac{G_t \cap P_t}{G_t \cup P_t} \tag{17}$$

In this expression, G_t is the set of the gold labels for tweet t, P_t is the set of the predicted labels for tweet t, and T is the set of tweets. Additionally, we also used the micro-averaged F-score and the macro-averaged F-score.

Let $\#_c(l)$ denote the number of samples correctly assigned to the label l, $\#_p(l)$ the number of samples assigned to l, and $\#(l)$ the number of actual samples in l. The micro-averaged F1-score is calculated as follows:

$$P_{micro} = \frac{\sum_{l \in L} \#_c(l)}{\sum_{l \in L} \#_p(l)} \tag{18}$$

$$R_{micro} = \frac{\sum_{l \in L} \#_c(l)}{\sum_{l \in L} \#(l)} \tag{19}$$

$$F1_{micro} = \frac{2 \times P_{micro} \times R_{micro}}{P_{micro} + R_{micro}} \tag{20}$$

Thus, P_{micro} is the micro-averaged precision score, and R_{micro} is the micro-averaged recall score.

Let P_l, R_l, and F_l denote the precision score, recall score, and the F1-score of the label l. The macro-averaged F1-score is calculated as follows:

$$P_l = \frac{\#_c(l)}{\#_p(l)} \tag{21}$$

$$R_l = \frac{\#_c(l)}{\#(l)} \tag{22}$$

$$F_l = \frac{2 \times P_l \times R_l}{P_l + R_l} \tag{23}$$

$$F1_{macro} = \frac{1}{|L|} \sum_{l \in L} F_l \tag{24}$$

4.5. Results

We submitted our system's predictions to the SemEval Task1:E-C challenge. The results were computed by the organizers on a golden test set, for which we did not have access to the golden labels.

Table 2 shows the results of our system and the results of the compared models (obtained from their associated papers). As can be observed from the reported results, our system achieved the top Jaccard index accuracy and macro-averaged F1 scores among all the state-of-the-art systems, with a competitive, but slightly lower score for the micro-average F1.

To get more insight about the performance of our system, we calculated the precision score, the recall score, and the F1 score of each label. The results of this analysis are shown in Figure 4. We found that our system gave the best performance on the "joy" label followed by the "anger", "fear", "disgust", and "optimism" labels. The obtained F1-score of these labels was above 70%. The worst performance was obtained on the "trust", "surprise", "anticipation", and "pessimism" labels. In most cases, our system gave a recall score higher than the precision score. It seems that the system was aggressive against the emotions "trust", "surprise", "anticipation", and "pessimism" (i.e., the system associated a low number of samples to these labels). This can be attributed to the low number of training examples for these emotions and to the Out-Of-Vocabulary (OOV) problem.

Table 2. Results of our system and state-of-the-art systems. The best values are in bold.

Model	Accuracy (Jaccard)	Micro F1	Macro F1
BNet(Our System)	**0.590**	0.692	**0.564**
SVM-Unigrams	0.442	0.57	0.443
Transformer	0.577	0.690	0.561
NTUA-SLP	0.588	**0.701**	0.528
TCS	0.582	0.693	0.530
PlusEmo2Vec	0.576	0.692	0.497

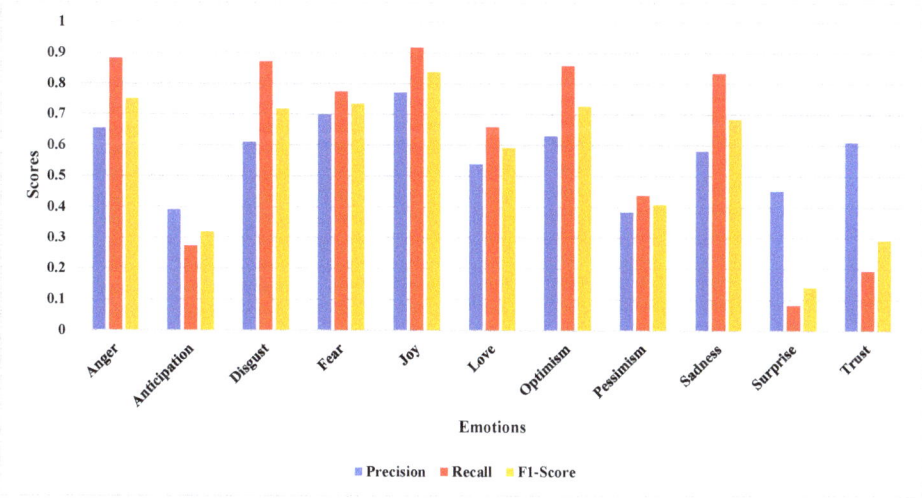

Figure 4. Performance analysis.

4.6. Attention Visualizations

We visualized the attention weights to get a better understanding of the performance of our system. The results are described in Figures 5–8, which show heat-maps of the attention weights on the top four example tweets from the validation set. The color intensity refers to the weight given to each word by the attention model. It represents the strength of the relationship between the word and the emotion, which reflects the importance of this word in the final prediction. We can see that the attention model gave the important weights to the common words, such as the stop words, in case the tweet was not assigned to the emotion; for example, the word "for" in Figure 5 and the word "this" in Figure 7 and the token "<user>" in Figure 8. Moreover, it also gives a high weight for the words and the emojis related to emotions (e.g., "cheering" and "awesome" for joy, "birthday" for love, etc.). An interesting observation is that when emojis were present, they were almost always selected as important if they were related to the emotion. For instance, we can see in Figure 7 that the sadness emotion relied heavily on the emoji. We also found that considering only one word to model the relation between the tweet and the emotions was not enough. In some cases, the emotion of a word may be flipped based on the context. For instance, consider the following tweet as an example: "When being #productive (doing the things that NEED to be done), #anxiety level decreases and #love level increases. #personalsexuality", the word "anxiety" is highly related to the emotion fear, but in this context, it shows optimism and trust emotions. However, our system misassociated this example with the fear emotion.

	im	clapping	and	cheering	for	both	teams	.	..1	..2
Anger	0.43	0.10	0.18	0.01	0.00	0.01	0.02	0.08	0.08	0.08
Anticipation	0.03	0.02	0.05	0.04	0.28	0.07	0.19	0.11	0.11	0.11
Disgust	0.15	0.08	0.26	0.01	0.01	0.02	0.04	0.14	0.14	0.14
Fear	0.01	0.01	0.02	0.00	0.23	0.01	0.00	0.24	0.24	0.24
Joy	0.01	0.07	0.00	0.74	0.13	0.04	0.01	0.00	0.00	0.00
Love	0.00	0.00	0.00	0.01	0.95	0.00	0.00	0.01	0.01	0.01
Optimism	0.01	0.01	0.02	0.13	0.50	0.18	0.09	0.02	0.02	0.02
Pessimism	0.01	0.01	0.02	0.01	0.30	0.01	0.03	0.20	0.20	0.20
Sadness	0.03	0.02	0.03	0.02	0.21	0.02	0.02	0.21	0.21	0.21
Surprise	0.00	0.00	0.00	0.00	1.00	0.00	0.00	0.00	0.00	0.00
Trust	0.00	0.00	0.00	0.00	1.00	0.00	0.00	0.00	0.00	0.00

Figure 5. Attention visualization example. Golden labels are {*joy, optimism*} and predicted labels are {*joy (0.91), optimism (0.51)*}.

	i	got	a	free	dr	.	pepper	from	the	vending	machine	awesome
Anger	0.02	0.03	0.28	0.04	0.01	0.12	0.07	0.04	0.33	0.02	0.04	0.01
Anticipation	0.28	0.05	0.04	0.05	0.27	0.07	0.03	0.01	0.02	0.01	0.05	0.02
Disgust	0.02	0.03	0.22	0.04	0.02	0.14	0.04	0.00	0.31	0.04	0.04	0.01
Fear	0.11	0.00	0.16	0.00	0.00	0.12	0.00	0.00	0.53	0.04	0.01	0.00
Joy	0.02	0.01	0.00	0.00	0.00	0.20	0.02	0.00	0.00	0.00	0.00	0.88
Love	0.70	0.01	0.01	0.01	0.00	0.01	0.00	0.00	0.04	0.00	0.00	0.16
Optimism	0.83	0.10	0.01	0.00	0.13	0.02	0.03	0.01	0.02	0.00	0.02	0.23
Pessimism	0.08	0.01	0.20	0.01	0.01	0.18	0.01	0.01	0.46	0.01	0.00	0.00
Sadness	0.07	0.02	0.19	0.01	0.01	0.19	0.01	0.04	0.34	0.01	0.02	0.01
Surprise	0.49	0.00	0.00	0.00	0.00	0.00	0.00	0.00	0.00	0.00	0.00	0.00
Trust	0.50	0.00	0.00	0.00	0.00	0.00	0.00	0.50	0.00	0.00	0.00	0.00

Figure 6. Attention visualization example. Golden labels are {*joy, surprise*} and predicted labels are {*joy (0.97), optimism (0.87)*}.

	can	not	believe	zain	starting	secondary	this	year	😊
Anger	0.25	0.13	0.02	0.02	0.02	0.06	0.49	0.00	0.02
Anticipation	0.09	0.03	0.06	0.17	0.34	0.17	0.04	0.08	0.02
Disgust	0.14	0.17	0.04	0.03	0.04	0.10	0.45	0.01	0.03
Fear	0.19	0.02	0.02	0.01	0.02	0.03	0.69	0.01	0.01
Joy	0.04	0.00	0.10	0.12	0.12	0.02	0.01	0.57	0.03
Love	0.13	0.01	0.08	0.14	0.01	0.00	0.21	0.08	0.34
Optimism	0.03	0.01	0.19	0.26	0.12	0.07	0.02	0.27	0.02
Pessimism	0.34	0.06	0.03	0.03	0.02	0.04	0.35	0.01	0.13
Sadness	0.01	0.02	0.01	0.01	0.00	0.01	0.03	0.00	0.91
Surprise	0.14	0.00	0.00	0.00	0.00	0.00	0.85	0.00	0.00
Trust	0.34	0.00	0.00	0.00	0.00	0.00	0.65	0.00	0.00

Figure 7. Attention visualization example. Golden labels are {*sadness, surprise*} and predicted labels are {*love (0.74), sadness (0.98)*}.

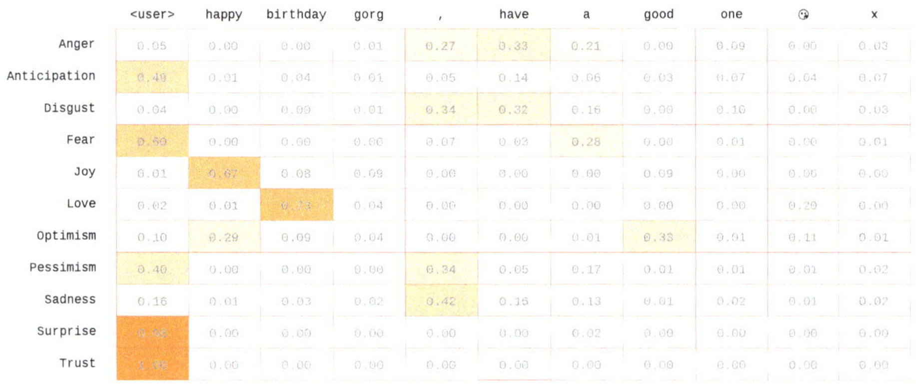

	<user>	happy	birthday	gorg	,	have	a	good	one	😊	x
Anger	0.05	0.00	0.00	0.01	0.27	0.33	0.21	0.00	0.09	0.00	0.03
Anticipation	0.49	0.01	0.04	0.01	0.05	0.14	0.06	0.03	0.07	0.04	0.07
Disgust	0.04	0.00	0.00	0.01	0.34	0.32	0.16	0.00	0.10	0.00	0.03
Fear	0.50	0.00	0.00	0.00	0.07	0.03	0.28	0.00	0.01	0.00	0.01
Joy	0.01	0.67	0.08	0.09	0.00	0.00	0.00	0.09	0.00	0.00	0.00
Love	0.02	0.01	0.73	0.04	0.00	0.00	0.00	0.00	0.00	0.20	0.00
Optimism	0.10	0.29	0.09	0.04	0.00	0.00	0.01	0.33	0.01	0.11	0.01
Pessimism	0.40	0.00	0.00	0.00	0.34	0.05	0.17	0.01	0.01	0.01	0.02
Sadness	0.16	0.01	0.03	0.02	0.42	0.16	0.13	0.01	0.02	0.01	0.02
Surprise		0.00	0.00	0.00	0.00	0.00	0.02	0.00	0.00	0.00	0.00
Trust		0.00	0.00	0.00	0.00	0.00	0.00	0.00	0.00	0.00	0.00

Figure 8. Attention visualization example. Golden labels are {*joy, love, optimism*} and predicted labels are {*joy* (0.98), *love* (0.91) *optimism* (0.95)}.

4.7. Correlation Analysis

Figure 9 shows the correlation analysis of emotion labels in the validation set. Each cell in the figure represents the correlation score of each pair of emotion labels. The reported values show exciting findings. Our system captured the relations among the emotion labels. The correlation scores of the predicted labels were almost identical to the ground-truth. There was an exception in the surprise and trust emotions. Our system was unsuccessful in capturing the relationships between these two emotions and the inputs or the other emotions. We attribute this apparent lack of correlation to the low number of training examples of these two emotions.

Moreover, there was always a positive correlation between related emotions such as "joy" and "optimism" (the score from the ground truth labels and from the predicted labels was 0.74). On the other side, we can see that there was a negative correlation between unlinked emotions like "anger" and "love". The scores were −0.27 and −0.3, respectively.

This result further strengthened our hypothesis that the proposed system was able to, implicitly, model the relationships between the emotion labels.

	Anger	Anticipation	Disgust	Fear	Joy	Love	Optimism	Pessimism	Sadness	Surprise	Trust
Anger	1	-0.15	0.7	-0.021	-0.521	-0.27	-0.45	0.055	0.15	-0.054	-0.16
Anticipation	-0.15	1	-0.15	-0.066	0.092	-0.05	0.14	-0.051	-0.12	0.15	0.11
Disgust	0.7	-0.15	1	0.024	-0.53	-0.29	-0.49	0.015	0.2	-0.068	-0.17
Fear	-0.021	-0.066	0.024	1	-0.24	-0.15	-0.19	0.076	0.0058	-0.064	-0.029
Joy	-0.521	0.092	-0.53	-0.24	1	0.4	0.58	-0.22	-0.35	0.049	0.13
Love	-0.27	-0.05	-0.29	-0.15	0.4	1	0.31	-0.14	-0.2	-0.069	0.11
Optimism	-0.45	0.14	-0.49	-0.19	0.58	0.31	1	-0.2	-0.28	0.011	0.24
Pessimism	0.055	-0.051	0.015	0.076	-0.22	-0.14	-0.2	1	0.34	-0.036	-0.081
Sadness	0.15	-0.12	0.2	0.0058	-0.35	-0.2	-0.28	0.34	1	-0.069	-0.12
Surprise	-0.054	0.15	-0.068	-0.064	0.049	-0.069	0.011	-0.036	-0.069	1	0.062
Trust	-0.16	0.11	-0.17	-0.029	0.13	0.11	0.24	-0.081	-0.12	0.062	1

(**a**) The ground-truth labels.

	Anger	Anticipation	Disgust	Fear	Joy	Love	Optimism	Pessimism	Sadness	Surprise	Trust
Anger	1	-0.14	0.78	0.054	-0.53	-0.3	-0.57	-0.038	0.24	0.04	-0.075
Anticipation	-0.14	1	-0.15	-0.064	0.21	-0.03	0.23	-0.044	-0.12	0.15	0.088
Disgust	0.78	-0.15	1	0.086	-0.54	-0.32	-0.59	0.015	0.31	0.04	-0.081
Fear	0.054	-0.064	0.086	1	-0.22	-0.12	-0.16	0.044	0.049	0.082	-0.039
Joy	-0.53	0.21	-0.54	-0.22	1	0.41	0.74	-0.18	-0.38	0.035	0.051
Love	-0.3	-0.03	-0.32	-0.12	0.41	1	0.46	-0.088	-0.21	0.077	0.023
Optimism	-0.57	0.23	-0.59	-0.16	0.74	0.46	1	-0.17	-0.4	0.04	0.065
Pessimism	-0.038	-0.044	0.015	0.044	-0.18	-0.088	-0.17	1	0.35	0.12	-0.027
Sadness	0.24	-0.12	0.31	0.049	-0.38	-0.21	-0.4	0.35	1	0.044	-0.048
Surprise	0.04	0.15	0.04	0.082	0.035	0.077	0.04	0.12	0.044	1	-0.0032
Trust	-0.075	0.088	-0.081	-0.039	0.051	0.023	0.065	-0.027	-0.048	-0.0032	1

(**b**) The predicted labels.

Figure 9. Correlation matrices of emotion labels of the development set.

5. Conclusions

In this work, we presented a new approach to the multi-label emotion classification task. First, we proposed a transformation method to transform the problem into a single binary classification problem. Afterwards, we developed a deep learning-based system to solve the transformed problem. The key component of our system was the embedding module, which used three embedding models and an attention function. Our system outperformed the state-of-the-art systems, achieving a Jaccard (i.e., multi-label accuracy) score of 0.59 on the challenging SemEval2018 Task 1:E-c multi-label emotion classification problem.

We found that the attention function can model the relationships between the input words and the labels, which helps to improve the system's performance. Moreover, we showed that our system is interpretable by visualizing the attention weights and analyzing them. However, some limitations have been identified. Our system does not model the relationships between the phrases and the labels. Phrases play a key role in determining the most appropriate set of emotions that must be assigned to a tweet. For instance, an emotion word that reflects "sadness" can be flipped in a negated phrase or context. Thus, in our future work, we plan to work on solving this drawback. One possible solution is to adapt the attention function to model the relationships between different n-gram tokens and labels. Structured attention networks [40] can also be adapted and used to address this issue.

Moreover, we plan to work on developing a non-aggressive system that performs robustly and equally on all the emotion labels by experimenting with different ideas like using data augmentation to enrich the training data or using transfer learning.

Author Contributions: Conceptualization, M.J. and A.M.; methodology, M.J.; software, M.J.; validation, M.J. and A.M.; formal analysis, M.J. and A.M.; investigation, M.J.; resources, M.J.; data curation, M.J.; writing, original draft preparation, M.J.; writing, review and editing, A.M.; visualization, M.J.; supervision, A.M.; funding acquisition, A.M.

Acknowledgments: The authors acknowledge the research support of URV(2018PFR-URV-B2 and Martí i Franqués PhD grant).

Conflicts of Interest: The authors declare no conflict of interest.

References

1. Chen, Y.; Zhou, Y.; Zhu, S.; Xu, H. Detecting Offensive Language in Social Media to Protect Adolescent Online Safety. In Proceedings of the 2012 International Conference on Privacy, Security, Risk and Trust (PASSAT), and 2012 International Conference on Social Computing (SocialCom), Amsterdam, The Netherlands, 3–5 September 2012; pp. 71–80.
2. Cherry, C.; Mohammad, S.M.; Bruijn, B. Binary Classifiers and Latent Sequence Models for Emotion Detection in Suicide Notes. *Biomed. Inform. Insights* **2012**, *5*, BII-S8933. [CrossRef] [PubMed]
3. Mohammad, S.M.; Zhu, X.; Kiritchenko, S.; Martin, J. Sentiment, Emotion, Purpose, and Style in Electoral Tweets. *Inf. Process. Manag.* **2015**, *51*, 480–499. [CrossRef]
4. Cambria, E. Affective computing and sentiment analysis. *IEEE Intell. Syst.* **2016** *31*, 102–107. [CrossRef]
5. Jabreel, M.; Moreno, A.; Huertas, A. Do Local Residents and Visitors Express the Same Sentiments on Destinations Through Social Media? In *Information and Communication Technologies in Tourism*; Springer: New York, NY, USA, 2017; pp. 655–668.
6. Yun, H.Y.; Lin, P.H.; Lin, R. Emotional Product Design and Perceived Brand Emotion. *Int. J. Adv. Psychol. IJAP* **2014**, *3*, 59–66. [CrossRef]
7. Mohammad, S.M. Sentiment Analysis: Detecting Valence, Emotions, and Other Affectual States from Text. In *Emotion Measurement*; Meiselman, H.L., Ed.; Woodhead Publishing: Cambridge, UK, 2016; pp. 201–237, ISBN 9780081005088.
8. Read, J.; Pfahringer, B.; Holmes, G.; Frank, E. Classifier Chains for Multi-label Classification. *Mach. Learn.* **2011**, *85*, 333. [CrossRef]
9. Robert, P. Emotions: A general Psychoevolutionary Theory. In *Approaches to Emotion*; Scherer, K.R., Ekman, P., Eds.; Psychology Press: London, UK, 2014; pp. 197–219.

10. Tsoumakas, G.; Katakis, I. Multi-label Classification: An Overview. *Int. J. Data Warehous. Min. IJDWM* **2007**, *3*, 1–13. [CrossRef]
11. Read, J. Scalable Multi-label Classification. Ph.D. Thesis, University of Waikato, Hamilton, New Zealand, 2010.
12. Mikolov, T.; Chen, K.; Corrado, G.; Dean, J. Efficient Estimation of Word Representations in Vector Space. *arXiv* **2013**, arXiv:1301.3781.
13. Schuster, M.; Paliwal, K.K. Bidirectional Recurrent Neural Networks. *IEEE Trans. Signal Process.* **1997**, *45*, 2673–2681. [CrossRef]
14. Chung, J.; Gulcehre, C.; Cho, K.; Bengio, Y. Empirical Evaluation of Gated Recurrent Neural Networks on Sequence Modeling. *arXiv* **2014**, arXiv:1412.3555.
15. Al-Molegi, A.; Jabreel, M.; Martínez-Ballesté, A. Move, Attend and Predict: An attention-based neural model for people's movement prediction. *Pattern Recognit. Lett.* **2018**, *112*, 34–40. [CrossRef]
16. Vaswani, A.; Shazeer, N.; Parmar, N.; Uszkoreit, J.; Jones, L.; Gomez, A.N.; Kaiser, Ł.; Polosukhin, I. Attention is All You Need. In Proceedings of the 31st Conference on Neural Information Processing Systems (NIPS 2017), Long Beach, CA, USA, 4–9 December 2017; pp. 5998–6008.
17. Zhang, M.L.; Zhou, Z.H. A review on Multi-label Learning Algorithms. *IEEE Trans. Knowl. Data Eng.* **2014**, *26*, 1819–1837. [CrossRef]
18. Zhang, M.L.; Zhang, K. Multi-label Learning by Exploiting Label Dependency. In Proceedings of the 16th ACM International Conference on Knowledge Discovery and Data Mining (SIGKDD), Washington, DC, USA, 25–28 July 2010; pp. 999–1008.
19. Schapire, R.E.; Singer, Y. BoosTexter: A boosting-Based System for Text Categorization. *Mach. Learn.* **2000** *39*, 135–168. [CrossRef]
20. Zhang, M.L.; Zhou, Z.H. ML-KNN: A Lazy Learning Approach to Multi-label Learning. *Pattern Recognit.* **2007**, *40*, 2038–2048. [CrossRef]
21. Clare, A.; King, R.D. Knowledge Discovery in Multi-label Phenotype Data. In Proceedings of the European Conference on Principles of Data Mining and Knowledge Discovery, Freiburg, Germany, 3–5 September 2001; pp. 42–53.
22. De Comite, F., Gilleron, R.; Tommasi, M. Learning Multi-label Alternating Decision Trees from Texts and Data. In Proceedings of the 3rd International Conference on Machine Learning and Data Mining in Pattern Recognition, Leipzig, Germany, 5–7 July 2003; Volume 2734, pp. 35–49.
23. Mencia, E.L.; Fürnkranz, J. Efficient Pairwise Multilabel Classification for Large-Scale Problems in the Legal Domain. In Proceedings of the Joint European Conference on Machine Learning and Knowledge Discovery in Databases, Antwerp, Belgium, 15–19 September 2008; Springer: Berlin/Heidelberg, Germany, 2008; pp. 50–65.
24. Cheng, W.; Hüllermeier, E. Combining Instance-Based Learning and Logistic Regression for Multilabel Classification. *Mach. Learn.* **2009**, *76*, 211–225. [CrossRef]
25. Godbole, S.; Sarawagi, S. Discriminative Methods for Multi-labeled Classification. In Proceedings of the Pacific-Asia Conference on Knowledge Discovery and Data Mining, Osaka, Japan, 20–23 May 2008; Springer: Berlin/Heidelberg, Germany, 2008; pp. 22–30.
26. Younes, Z.; Abdallah, F.; Denoeux, T.; Snoussi, H. A Dependent Multilabel Classification Method Derived From the k-Nearest Neighbor Rule. *J. Adv. Signal Process.* **2011**, *1*, 645964. [CrossRef]
27. Yan, R.; Tesic, J.; Smith, J.R. Model-Shared Subspace Boosting for Multi-label Classification. In Proceedings of the 13th ACM SIGKDD, San Jose, CA, USA, 12–15 August 2007; pp. 834–843.
28. Jabreel, M.; Moreno, A. SentiRich: Sentiment Analysis of Tweets Based on a Rich Set of Features. In *Artificial Intelligence Research and Development*; Nebot, Á., Binefa, X., López de Mántaras, R., Eds.; IOS Press: Amsterdam, The Netherlands, 2016; Volume 288, pp. 137–146, ISBN 978-1-61499-695-8.
29. Jabreel, M.; Moreno, A. SiTAKA at SemEval-2017 Task 4: Sentiment Analysis in Twitter Based on a Rich Set of Features. In Proceedings of the 11th International Workshop on Semantic Evaluation (SemEval-2017), Vancouver, BC, Canada, 3–4 August 2017; pp. 694–699.
30. Mohammed, S., M.; Bravo-Marquez, F.; Salameh, M.; Kiritchenko,S. Semeval-2018 task 1: Affect in Tweets. In Proceedings of the 12th International Workshop on Semantic Evaluation, New Orleans, LA, USA, 5–6 June 2018; pp. 1–17.
31. LeCun, Y.; Bengio, Y.; Hinton, G. Deep Learning. *Nature* **2015** *521*, 436–444. [CrossRef]

32. Tang, D.; Qin, B.; Liu, T. Deep Learning for Sentiment Analysis: Successful Approaches and Future Challenges. *Wiley Interdiscip. Rev. Data Min. Knowl. Discov.* **2015** 5, 292–303. [CrossRef]
33. Baziotis, C.; Athanasiou, N.; Chronopoulou, A.; Kolovou, A.; Paraskevopoulos, G.; Ellinas, N.; Narayanan, S.; Potamianos, A. NTUA-SLP at SemEval-2018 Task 1: Predicting Affective Content in Tweets with Deep Attentive RNNs and Transfer Learning. In Proceedings of the 12th International Workshop on Semantic Evaluation, New Orleans, LA, USA, 5–6 June 2018; pp. 245–255.
34. Meisheri, H.; Dey, L. TCS Research at Semeval2018 Task 1: Learning Robust Representations using Multi-Attention Architecture. In Proceedings of the 12th International Workshop on Semantic Evaluation, New Orleans, LA, USA, 5–6 June 2018; pp. 291–299.
35. Park, J.H.; Xu, P.; Fung, P. PlusEmo2Vec at SemEval-2018 Task 1: Exploiting Emotion Knowledge from Emoji and #hashtags. In Proceedings of the 12th International Workshop on Semantic Evaluation, New Orleans, LA, USA, 5–6 June 2018; pp. 264–272.
36. Kingma, D.P.; Ba, J. Adam: A Method for Stochastic Optimization. *arXiv* **2014**, arXiv:1412.6980.
37. James, B.; Yamins, D.; Cox, D.D. Hyperopt: A python library for optimizing the hyperparameters of machine learning algorithms. In Proceedings of the 12th Python in Science Conference, Austin, TX, USA, 24–29 June 2013.
38. Mohammad, S.; Kiritchenko, S. Understanding emotions: A dataset of tweets to study interactions between affect categories. In Proceedings of the Eleventh International Conference on Language Resources and Evaluation, Miyazaki, Japan, 7–12 May 2018.
39. Kant, N.; Puri, R.; Yakovenko, N.; Catanzaro, B. Practical Text Classification With Large Pre-Trained Language Models. *arXiv* **2018**, arXiv:1812.01207.
40. Kim, Y.; Denton, C.; Hoang, L.; Rush, A.M. Structured Attention Networks. *arXiv* **2017**, arXiv:1702.00887.

© 2019 by the authors. Licensee MDPI, Basel, Switzerland. This article is an open access article distributed under the terms and conditions of the Creative Commons Attribution (CC BY) license (http://creativecommons.org/licenses/by/4.0/).

Article

Using Social Media to Identify Consumers' Sentiments towards Attributes of Health Insurance during Enrollment Season

Eline M. van den Broek-Altenburg * and Adam J. Atherly

Center for Health Services Research, The Larner College of Medicine, University of Vermont, Burlington, VT 05405, USA; adam.atherly@med.uvm.edu
* Correspondence: eline.altenburg@med.uvm.edu; Tel.: +1-(802)-656-2722; Fax: +1-(802)-656-8577

Received: 27 April 2019; Accepted: 13 May 2019; Published: 17 May 2019

Featured Application: The health insurance choice literature has found that financial considerations, such as premiums, deductible, and maximum out-of-pocket spending caps, are important to consumers. But these financial factors are just part of the cost-benefit trade-off consumers make. Publicly available datasets often do not include these other factors. Researchers in other fields have increasingly used web data from social media platforms, such as Twitter and search engines to analyze consumer behavior using Natural Language Processing. NLP combines machine learning, computational linguistics, and computer science, to understand natural language including consumer's sentiments, attitudes, and emotions from social media. This study is among the first to use natural language from an online platform to analyze sentiments when consumers are discussing health insurance. By clarifying what the expressed attitudes or sentiments are, we get an idea of what variables we may want to include in future studies of health insurance choice.

Abstract: This study aims to identify sentiments that consumers have about health insurance by analyzing what they discuss on Twitter. The objective was to use sentiment analysis to identify attitudes consumers express towards health insurance and health care providers. We used an Application Programming Interface to gather tweets from Twitter with the words "health insurance" or "health plan" during health insurance enrollment season in the United States in 2016–2017. Word association was used to find words associated with "premium," "access," "network," and "switch." Sentiment analysis established which specific emotions were associated with insurance and medical providers, using the NRC Emotion Lexicon, identifying emotions. We identified that provider networks, prescription drug benefits, political preferences, and norms of other consumers matter. Consumers trust medical providers but they fear unexpected health events. The results suggest that there is a need for different algorithms to help consumers find the plans they want and need. Consumers buying health insurance in the Affordable Care Act marketplaces in the United States choose lower-cost plans with limited benefits, but at the same time express fear about unexpected health events and unanticipated costs. If we better understand the origin of the sentiments that drive consumers, we may be able to help them better navigate insurance plan options and insurers can better respond to their needs.

Keywords: social media; Twitter; text mining; sentiment analysis; word association; health insurance; provider networks

1. Introduction

In the Affordable Care Act health insurance marketplaces in the United States (USA), consumers are mandated to choose a health insurance plan. Plans may differ by premiums, benefits, and other plan attributes, such as the network of providers or how tightly managed the plan is. Consumers ideally pick the best combination of plan attributes, switching plans if necessary.

The health insurance choice literature has found that financial considerations, such as premiums, deductibles, and maximum out-of-pocket spending caps, are indeed important to consumers [1–5]. However, these considerations are just part of the cost-benefit trade-off consumers make. Surveys and discrete choice experiments suggest that other plan attributes, such as choice of personal doctors [6,7], continuity of care [8–11], or how "tightly managed" the plan is [4], also have an effect on consumers' choices. Information about quality of service or other aspects of care delivery may also play a role [12]. The more we know about the trade-offs consumers make and what factors play a role in insurance choice, the better we can predict or anticipate future choices.

This study identifies sentiments that consumers have when discussing health insurance in the USA by using an alternative data source: Twitter. Twitter has grown exponentially in recent years and computer and data scientists have learned how to extract information from the 328 million monthly active Twitter users, 70 million of whom live in the USA [13], Every second, on average, around 6000 tweets are sent via Twitter, which corresponds to 500 million tweets per day and around 200 billion tweets per year [14].

Twitter's "tweets," which were at the time of our study limited to 140 characters, have been shown to have surprising predictive power. Numerous studies across different academic fields have used Twitter as a tool for forecasting or prediction. Researchers in industrial organization and marketing have used Twitter data to analyze what consumers want and need. In fields like finance and macroeconomics, text from social media has been used to make predictions about the stock market [15–17], oil [18], sales [19], and unemployment rates [20,21], or as a surveillance tool to track messages related to security breaches [22]. In the political arena, Twitter has been used to predict the outcome of elections or to poll political sentiment [23–25]. It has been suggested that analysis of social media data more accurately predicted Trump's win than election polls [26].

More recently, text mining of web content has been used in the context of public health. Twitter data have been used to evaluate health care quality, poll reactions to health policy reforms and in various other public health contexts. Additionally, researchers have used text from Twitter for influenza surveillance [27–31]. For example, an analysis of three million tweets between May and December 2009 showed that the 2009 H1N1 flu outbreak could have been identified on Twitter one week before it emerged in official records from general practitioner reports. Researchers at the Department of Computer Science at Johns Hopkins University created a model for Twitter that groups symptoms and treatments into latent ailments.

Other examples include using tweets to compute the average happiness of cancer patients for each cancer diagnosis [32], to measure patient-perceived quality of care in hospitals [33], and to predict asthma prevalence by combining Twitter data with other data sources [34]. The latter study provides evidence that monitoring asthma-related tweets may provide real-time information that can be used to predict outcomes from traditional surveys.

Some recent studies have used web data from search engines such as Google to analyze consumer behavior in health insurance. One study examined factors associated with health insurance-related Google searches during the first open enrollment period [35]. The authors found that search volumes were associated with local uninsured rates. Another study used text data from Twitter to identify consumers' sentiments to predict insurance enrollment [36].

A number of studies have used Twitter data in a similar context to this study. Beyond the health insurance studies mentioned above, Twitter has also been used to assess public opinion about the Affordable Care Act over time: a study found substantial spikes in the volume of Affordable Care Act-related tweets in response to key events in the law's implementation [37].

The aim of this study is to identify sentiments that consumers express on Twitter when they discuss health insurance. The first objective of this paper is to identify words that are associated with the word "switch" in the tweets. In the context of tweets gathered on the search "health insurance," we assume that "switch" is related to health insurance at least some of the time. The second objective of this paper is to identify what attitudes or sentiments consumers have when communicating about health insurance in online social networks. The study is hypothesis-generating: gaining insights into the words consumers use when they communicate about health insurance on an online social network may lead to better-informed theory regarding health plan choices. By clarifying what the expressed attitudes or sentiments are, we may find variables we can include in future studies and we may be able to generate testable hypotheses.

2. Materials and Methods

2.1. Data

Using an Application Programming Interface (API), we gathered tweets from the Twitter server with the words "health insurance," "health plan," "health provider" or "doctor" in them during open enrollment period from 1 November 2016 until 31 January 2017. This is the yearly period when U.S. citizens can enroll in or switch a health insurance plan. Beyond this timeframe, they have to stay with the plan they have. API is code that allows two software programs, in our case Twitter and Python 3.6, to communicate with each other. With the API, Python authenticated, requested, and received the data from the Twitter server. The words "health insurance" and "health plan" generated approximately one tweet every 3 s, adding up to 28,800 per day; 892,800 per month; and 2,678,400 total tweets during the ACA open enrollment season for 2017.

We used the API to create a body of text, called "VCorpus," in R 3.4. At each index of the "VCorpus object," there is a PlainTextDocument object, which is essentially a list that contains the actual text data of the tweet, as well as some corresponding metadata such as the location from which the tweet was sent, the date, and other elements. In other words, the tweets were gathered in one text document and pre-processed for analysis. This pre-processing gets rid of punctuation, hashtags, and retweets, strips white space, and removes stop words and custom terms so that they are now represented as lemmatized plain words. To illustrate, the tweet text "Obama care is a joke fr. My health care plan is just not affordable no more. Cheaper to pay the penalty I guess" was changed to: "obama care joke fr health care plan just affordable cheaper pay penalty guess" after pre-processing.

The most important way that text differs from more typical data sources is that text is naturally high-dimensional, which makes analysis difficult, often referred to as the "curse of dimensionality." For example, suppose that a sample of tweets, each of which is 20 words long, and that each word is drawn from a vocabulary of 2000 possible words. It follows that the unique representation of these tweets has a very high dimensionality, with 40,000 data columns.

To reduce dimensionality, we use the "bag of words" (BoW) model. The BoW model, also known as a vector space model, reduces dimensionality by simplifying the representation of the words used in natural language processing and information retrieval. In this model, a text document (such as a tweet) is represented as the bag (multiset) of its words, disregarding grammar and word order [38].

Subsequently, our "bag of words" model learned a vocabulary from the millions of tweets and then modeled each tweet by counting the number of times each word appears in the tweet [38]. Through automatic text categorization, we extracted features or "token sets" from the text by representing the tweets by the words that occur in it.

To explain how we converted text to numeric data, here is an example. Sentence 1: "The health insurance plan is too expensive to cover my health needs"; Sentence 2: "The health insurance company offers an expensive health plan." We can see that, from these two sentences, our vocabulary is: {The, health, insurance, plan, is, too, expensive, to, cover, my, needs, company, offers, an}. To get the bags of words, the number of times each word occurs was counted in each sentence. In Sentence 1," health" appears twice, and the other words each appear once, so the feature vector for Sentence 1 is: {1, 2, 1, 1, 1, 1, 1, 1, 1, 1, 1, 0, 0, 0} and Sentence 2: {1, 2, 1, 0, 0, 0, 0, 0, 0, 0, 0, 1, 1, 1} We created a Term-Document matrix (TDM) where each row is a 1/0 representation of whether a single word is contained within the tweet and every column is a tweet. We then removed terms with a sparse factor of less than 0.001. These are the terms that occur less than 0.01% of times in a tweet. The resulting matrix contained 1354 words.

2.2. Analytic Approach

To find which words are associated with switching, we used Word Association: a function that calculates the association of a word with another word in the TDM. We used the findAssocs() in R to calculate the association of a word with every other word in the TDM. The output scores range from 0 to 1, where a score of 1 means that two words always appear together, and a score of 0 means that they never appear together. To find associations, we set a minimum of 0.05, meaning that the program would look for all words that were associated in one tweet (that has a maximum of 140 characters) with "premium" at least 5% of the tweets. Since we were interested in attitudes to plan attributes, we tested three attributes of health insurance plans: "premium," "access," and "network." We also looked for all the words that were associated with the word "switch" at least 5% of the time. We chose "premium" because we know from the literature that the premium matters when consumers buy health insurance, as well as "access" to doctors. Since we were particularly interested in whether provider networks, which refers to insurance coverage of doctors in-network, matters when consumers discuss health insurance, we also looked at "network."

To identify sentiments in the tweets, we use methods of classification. We used sentiment lexicons, a dictionary-based approach, which depends on finding opinion seed words, and then searches the dictionary for their synonyms. While various sentiment lexicons all have their advantages and disadvantages in the context of topic-specific subjectivity scores, interpretation, and completeness, the choice for a specific sentiment lexicon is context-specific. We used the NRC Emotion Lexicon (NRC Emolex) [39], which classifies words in a binary yes/no for classes of attitude "positive" and "negative"; and for classes of emotion: anger, anticipation, disgust, fear, joy, sadness, surprise, and trust. We wanted to not only find out the overall sentiment of tweets, but also what specific emotion they embodied, and identify which words represented those emotions.

3. Results

3.1. Word Association

Figure 1 shows that the most common word consumers used in combination with the word "switch" was the word "bait" (0.31), meaning that in 31% of tweets with the word "switch," the word "bait" was also used. In 8% of the tweets that use the word "switch," the word "premium" is also used.

This suggests that insurance was often described as a bait and switch, such as in this example tweet: "The healthcare bait-and-switch. In network hospital, out of network doctor." This was followed by "lockedin" and "rxrights." "Rxrights" refers RxRights.org, which serves as a forum for individuals to share experiences and voice opinions regarding the need for affordable prescription drugs. It is an example of how "switch" could be used in a different context than insurance, such as in this tweet: "In the USA, this is how we are forced to switch insulins without any regard to our health or to doctors' orders." The next most common word associated with switch was "network." Networks were used in tweets about switching, such as in this example: "Gods blessings is like health insurance, you have in network benefits and out of network benefits, but in network is way better of course." There were tweets discussing the role of provider networks in insurance such as this one: "$252 for a doctor visit. I wasn't even there for 20 minutes. Thanks insurance for not letting me know the doc was no longer in my network."

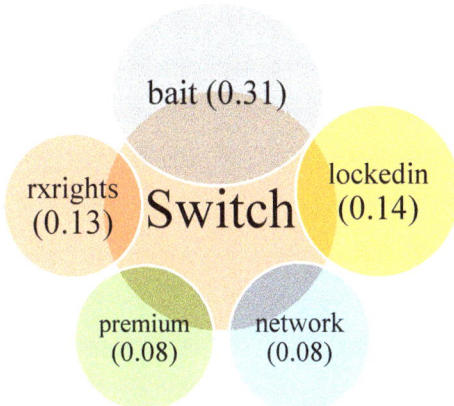

Figure 1. Word associations for plan attributes and switching among health insurance tweets.

"Network" was associated with the word switch as often as "premium," which was 0.08, meaning that 8% of tweets that had the word "switch" in them also contained the work "premium." Consumers expressed concerns about premiums, deductibles, and co-pays, such as in this example: "Dude, as expensive as my insurance is. Copays, premiums, etc., I might as well not even have it. Costs 2 much to use."

3.2. Sentiment Analysis

The results of the sentiment analysis showed that two emotions prevailed in tweets during enrollment season: "trust" and "fear" (Table 1). The emotion "trust" was the primary driver of positive sentiment expressed in tweets, while the emotion "fear" was the primary driver of negative sentiments and accounted for the slightly negative overall sentiment. Trust was expressed in the context of doctors, nurses, and other medical providers. Here is an example of a tweet discussing this trustworthy role: "Patients value their family doctor at the bedside when gravely ill. Healing presence is so powerful."

Table 1. Positive and negative emotions expressed on Twitter, by the words used in the tweets.

Positive	Negative
Doctor (trust)	Pressure (negative)
Physician (trust)	Die (fear, sadness)
Hospital (trust)	Emergency (fear, anger, disgust, sadness)
Nurse (trust)	Disease (fear, anger, disgust, sadness)
Plan (anticipation)	Pain, surgery (fear, sadness)
Save money (joy)	Miscarriage (fear, sadness)
Choice (positive)	Cancer (fear, anger, disgust, sadness)

Note: In brackets are the emotions that the NRC Emolex associated with the word used in the tweets.

In a tweet like this, the NRC Emolex classified the word "value" as positive and associated with "trust," while the word "healing" is classified as positive and associated with the emotions of anticipation, joy, and trust. Another tweet referred to the importance of continuity of care: "Seeing anxiety in culture from lack of relationship with continuity of fam Doctor." The word "anxiety" is classified by the NRC Emolex as negative and associated with the emotions of fear, anticipation, and sadness. In this way, the words used in tweets are given a subjectivity score and common ones are reported in Table 1.

Fear was conveyed in Tweets about medical events such as "lose," "disease," "emergency," "surgery," "cancer," and unanticipated costs. Consumers expressed both negative and positive sentiments about choice, but the NRC Emolex could not specify the exact emotion. For example, one tweet stated: "I hate choice so much that I've essentially disregarded my doctor because I've had such a low care of my own wellbeing." It also identified "pressure" as being negative but could not specify what kind of emotion consumers expressed. Overall, the sentiment of consumers toward health insurance was slightly negative, although most sentiments were not at either extreme (Figure 2).

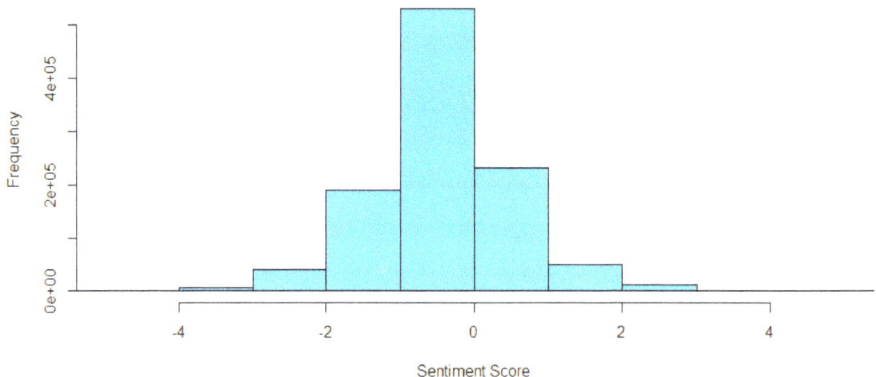

Figure 2. Sentiments expressed in health insurance-related tweets in 2016-2017 Affordable Care Act enrollment season.

The figure shows how the words used in the tweets were classified: either positive or negative. It follows from the histogram that most words were classified as "slightly negative" (-1) and few words were classified as either extremely negative or extremely positive.

To understand what attitudes consumers expressed regarding specific attributes of health plans, we examined "premium," "access," and "network." Table 2 illustrates that consumers used the insurance attribute "premium" most often in combination with words like "increase" or "relief."

Table 2. Words associated with premium, access, and network.

Term in Tweets	Words Used	Association
Premium	Increase	0.17
	relief	0.16
	Obamacare	0.08
Access	No-copay	0.74
	ppact (planned parenthood), birth, women	0.66, 0.51, 0.40
	Affordable Care Act	0.15
Network	narrow	0.23
	providers	0.16
	covering	0.08
Switch	Bait	0.31
	lockedin	0.14
	rxrights	0.13
	premium	0.08
	network	0.08

Access was associated most of the time with "nocopay," suggesting that consumers who care about access also care about copays. The attribute "network" was associated with "narrow" 23% of the time, and with "providers" 16% of the time, suggesting that many consumers talk about narrow network plans when discussing health insurance.

4. Discussion

In this study, we used text data from Twitter to identify attitudes that consumers express towards health insurance, plan attributes, and health care providers. The health insurance choice literature focuses primarily on well-defined features of plans that are easily observable from administrative data such as benefit design, co-insurance rates, and deductibles. Previous studies found that financial considerations, such as premiums, deductibles, and maximum out-of-pocket spending caps, are important to consumers. This study reinforces some results from previous research. The sensitivity of consumers to higher premiums that our study finds is well documented in other literature. The role of provider networks has been debated recently—our study reinforces the importance of the networks to consumers.

There are limitations associated with the bag of words approach that we used. The main disadvantage is that it severely limits the context of the tweet and loses the order of specific information. Also, it requires supervised machine learning, which entails modeling linguistic knowledge through the use of dictionaries containing words that are tagged with their semantic orientation [39]. This means that we used an existing data dictionary, but we accept the classification of the English words to identify emotions.

It is a challenge to capture the essential meaning of a tweet in a machine-understandable format [40] as issues like short length, informal words, misspellings, and unusual grammar make it difficult to obtain a good representation to capture these text aspects. More recently, there has been a paradigm shift in machine learning towards using distributed representations for words [41] and sentences [42,43]. Most studies analyzing tweets have not been able to use a fully unsupervised approach for message-level and phrase-level sentiment analysis of tweets. The advantage to such an approach would have been, for example, that we would have been able to convey emotions in the same manner in the tweets as in newspaper articles or blogs, reviews, or other types of user-generated content. In everyday life, we rely on context to interpret a piece of text or comment, so with bag of words it is harder to capture context as it merely focuses on isolated words or term frequencies.

We do not know how the demographics of people tweeting about health insurance compares to the Affordable Care Act marketplace population, the Medicaid expansion population, and the uninsured. Tweets contain macro data about the user, but it is limited to whatever information the user decides to give. In practice, a small percentage of users provide personal information such as gender or age. We do have some information about location, but we lack this information for a substantial part of the sample and the level of information (city, state) differs by user.

Another limitation is that tweets have social network-oriented properties, and therefore we believe that a good representation of our tweets should also capture social aspects. A social network analysis is beyond the scope of this study, as well as conversational science approach looking at how a comment is influenced by the previous one.

5. Conclusions

This study suggests that other, non-financial factors, might be important in the choice of health insurance plan, such as the sentiments that consumers have. The discussion of "fear" in relation to health insurance plan choice may seem unusual; however, the basic economic model for health insurance posits risk aversion as a key motivator for individuals to buy coverage. In some sense, "fear" is simply "risk-averse" in the vernacular. In another sense, however, this study provides specificity about the nature of the risk aversion and suggests that consumers lack confidence in their choices and express fear towards adverse health events and unanticipated costs.

If we better understand the origin of the fear and other sentiments that drive consumers, we may be able to help them to better navigate insurance plan options and insurers can make sure to better respond to their needs. Additionally, plan finders often provide consumers with actuarial spending estimates for "average" consumers; our study suggests that the average outcome is not the outcome of interest to consumers. Even though some plan finders include individual-specific calculations [44], insurance companies may want to react to consumers' sentiments in addition to financial considerations. Consumers are concerned about an unusual event—cancer, accident, surgery, or disease—and whether they can afford care when death is a real possibility. Plan finders could be reconfigured to give coverage data for consumers experiencing these extreme health events.

Text mining is an important addition to research in this area because of the sheer volume of information and the possibility of looking quantitatively at formative data. In social science in general, and public health research in particular, a common practice is to rely on small convenience samples to generate formative data that are qualitative to generate hypotheses. Potential testable hypotheses that were generated from the analysis may include "Provider networks are associated with health plan switching" or "Sentiments expressed on Twitter predict health insurance choice." Where qualitative research usually involves very small samples, text data can yield similar insights with substantially larger sample sizes. This study illustrates that we can use another, perhaps equally effective, advanced method and data to generate testable hypotheses.

Author Contributions: Both authors made substantial contributions to the conception and design of the work. E.M.v.d.B.-A. was responsible for the acquisition, analysis, and interpretation of data; and for drafting the work. A.J.A. played an important role in the interpretation of the data and results and substantively revised the manuscript. Both authors have approved the submitted version and agree to be personally accountable for their own contributions and for ensuring that questions related to the accuracy or integrity of any part of the work are appropriately investigated, resolved, and documented in the literature.

Conflicts of Interest: The authors declare no conflict of interest.

References

1. Atherly, A.; Dowd, B.E.; Feldman, R. The effect of benefits, premiums, and health risk on health plan choice in the Medicare program. *Health Serv. Res.* **2004**, *39*, 847–864. [CrossRef] [PubMed]
2. Buchmueller, T.C. *Consumer Demand for Health Insurance*; NBER Reporter, National Bureau of Economic Research, Inc.: Cambridge, MA, USA, 2006; pp. 10–13.
3. Short, P.F.; Taylor, A.K. Premiums, benefits, and employee choice of health insurance options. *J. Health Econ.* **1989**, *8*, 293–311. [CrossRef]
4. Trujillo, A.J.; Ruiz, F.; Bridges, J.F.P.; Amaya, J.L.; Buttorff, C.; Quiroga, A.M. Understanding consumer preferences in the context of managed competition. *Appl. Health Econ. Health Policy* **2012**, *10*, 99–111. [CrossRef]
5. Blumberg, L.J.; Long, S.K.; Kenney, G.M.; Goin, D. *Factors Influencing Health Plan Choice among the Marketplace Target Population on the Eve of Health Reform*; Urban Institute: Washington, DC, USA, 2013.
6. Altman, D. What new data tells us about doctors choice. *The Wall Street Journal*. 4 February 2016. Available online: https://blogs.wsj.com/washwire/2016/02/04/what-new-data-tell-us-about-doctor-choice/ (accessed on 28 September 2017).
7. Stokes, T.; Tarrant, C.; Mainous, A.G.; Schers, H.; Freeman, G.; Baker, R. Continuity of care: Is the personal doctor still important? A survey of general practitioners and family physicians in England and Wales, the United States, and The Netherlands. *Ann. Fam. Med.* **2005**, *3*, 353–359. [CrossRef] [PubMed]
8. Guthrie, B.; Saultz, J.W.; Freeman, G.K.; Haggerty, J.L. Continuity of care matters. *BMJ Br. Med. J.* **2008**, *337*, a867. [CrossRef]
9. Turner, D.; Tarrant, C.; Windridge, K.; Bryan, S.; Boulton, M.; Freeman, G.; Baker, R. Do patients value continuity of care in general practice? An investigation using stated preference discrete choice experiments. *J. Health Serv. Res. Policy* **2007**, *12*, 132–137. [CrossRef]
10. Higuera, L.; Carlin, C.S.; Dowd, B. Narrow provider networks and willingness to pay for continuity of care and network breadth. *J. Health Econ.* **2018**, *60*, 90–97. [CrossRef]
11. Mainous, A.G.; Goodwin, M.A.; Stange, K.C. Patient-physician shared experiences and value patients place on continuity of care. *Ann. Fam. Med.* **2004**, *2*, 452–454. [CrossRef]
12. Enthoven, A.; Kronick, R. Competition 101: Managing demand to get quality care. *Bus. Health* **1988**, *5*, 38–40.
13. Statistica. Number of Monthly Active Twitter Users in the United States from 1st Quarter 2010 to 1st Quarter 2017 (in Millions). 2017. Available online: https://www.statista.com/statistics/274564/monthly-active-twitter-users-in-the-united-states/ (accessed on 22 July 2017).

14. Stats IL. Twitter Usage Statistics. 2017. Available online: http://www.internetlivestats.com/twitter-statistics/ (accessed on 22 July 2017).
15. Ben-Ami, Z.; Feldman, R.; Rosenfeld, B. Using multi-view learning to improve detection of investor sentiments on twitter. *Computación y Sistemas* **2014**, *18*, 477–490. [CrossRef]
16. Bing, L.; Chan, K.C.; Ou, C. Public sentiment analysis in Twitter data for prediction of a company's stock price movements. In Proceedings of the 2014 IEEE 11th International Conference on e-Business Engineering (ICEBE), Guangzhou, China, 5–7 November 2014.
17. Chen, R.; Lazer, M. Sentiment analysis of twitter feeds for the prediction of stock market movement. *Stanf. Edu. Retrieved* **2013**, *25*, 2013.
18. Rao, T.; Srivastava, S. *Using Twitter Sentiments and Search Volumes Index to Predict Oil, Gold, Forex and Markets Indices*; Delhi Institutional Repository: Delhi, India, 2012.
19. Dijkman, R.; Ipeirotis, P.; Aertsen, F.; van Helden, R. Using twitter to predict sales: A case study. *arXiv* **2015**, arXiv:150304599.
20. Antenucci, D.; Cafarella, M.; Levenstein, M.; Ré, C.; Shapiro, M.D. *Using Social Media to Measure Labor Market Flows*; National Bureau of Economic Research: Cambridge, MA, USA, 2014.
21. Llorente, A.; Garcia-Herranz, M.; Cebrian, M.; Moro, E. Social media fingerprints of unemployment. *PLoS ONE* **2015**, *10*, e0128692. [CrossRef]
22. Hao, J.; Hao, J.; Dai, H.; Dai, H. Social media content and sentiment analysis on consumer security breaches. *J. Financ. Crime* **2016**, *23*, 855–869. [CrossRef]
23. Bermingham, A.; Smeaton, A.F. On using Twitter to monitor political sentiment and predict election results. In Proceedings of the Workshop at the International Joint Conference for Natural Language Processing (IJCNLP), Chiang Mai, Thailand, 13 November 2011.
24. Tumasjan, A.; Sprenger, T.O.; Sandner, P.G.; Welpe, I.M. Predicting elections with twitter: What 140 characters reveal about political sentiment. *ICWSM* **2010**, *10*, 178–185.
25. Tumasjan, A.; Sprenger, T.O.; Sandner, P.G.; Welpe, I.M. Election forecasts with Twitter: How 140 characters reflect the political landscape. *Soc. Sci. Comput. Rev.* **2011**, *29*, 402–418. [CrossRef]
26. Perez, S. Analysis of Social Media Did a Better Job at Predicting Trump's win than the Polls. Tech Crunch. 2016. Available online: https://techcrunch.com/2016/11/10/social-media-did-a-better-job-at-predicting-trumps-win-than-the-polls/ (accessed on 24 July 2017).
27. Broniatowski, D.A.; Paul, M.J.; Dredze, M. National and local influenza surveillance through Twitter: An analysis of the 2012–2013 influenza epidemic. *PLoS ONE* **2013**, *8*, e83672. [CrossRef] [PubMed]
28. Culotta, A. Towards detecting influenza epidemics by analyzing Twitter messages. In Proceedings of the First Workshop on Social Media Analytics, Washington, DC, USA, 25–28 July 2010.
29. Lamb, A.; Paul, M.J.; Dredze, M. Separating fact from fear: tracking flu infections on Twitter. In Proceedings of the North American Chapter of the Association for Computational Linguistics—Human Language Technologies (NAACL HLT) 2013 Conference, Atlanta, GA, USA, 9–14 June 2013.
30. Lampos, V.; De Bie, T.; Cristianini, N. Flu detector-tracking epidemics on Twitter. In Proceedings of the Joint European Conference on Machine Learning and Knowledge Discovery in Databases, Barcelona, Spain, 20–24 September 2010; pp. 599–602.
31. Signorini, A.; Segre, A.M.; Polgreen, P.M. The use of Twitter to track levels of disease activity and public concern in the US during the influenza A H1N1 pandemic. *PLoS ONE* **2011**, *6*, e19467. [CrossRef] [PubMed]
32. Crannell, W.C.; Clark, E.; Jones, C.; James, T.A.; Moore, J. A pattern-matched Twitter analysis of US cancer-patient sentiments. *J. Surg. Res.* **2016**, *206*, 536–542. [CrossRef]
33. Hawkins, J.B.; Brownstein, J.S.; Tuli, G.; Runels, T.; Broecker, K.; Nsoesie, E.O.; McIver, D.J.; Rozenblum, R.; Wright, A.; Bourgeois, F.T.; et al. Measuring patient-perceived quality of care in US hospitals using Twitter. *BMJ Qual. Saf.* **2016**, *25*, 404–413. [CrossRef]
34. Dai, H.; Lee, B.R.; Hao, J. Predicting asthma prevalence by linking social media data and traditional surveys. *ANNALS Am. Acad. Polit. Soc. Sci.* **2017**, *669*, 75–92. [CrossRef]
35. Gollust, S.E.; Qin, X.; Wilcock, A.D.; Baum, L.M.; Barry, C.L.; Niederdeppe, J.; Fowler, E.F.; Karaca-Mandic, P. Search and you shall find: Geographic characteristics associated with google searches during the affordable care act's first enrollment period. *Med. Care Res. Rev.* **2016**. [CrossRef] [PubMed]
36. Wong, C.A.; Sap, M.; Schwartz, A.; Town, R.; Baker, T.; Ungar, L.; Merchant, R.M. Twitter sentiment predicts affordable care act marketplace enrollment. *J. Med. Internet Res.* **2015**, *17*, e51. [CrossRef] [PubMed]

37. Davis, M.A.; Zheng, K.; Liu, Y.; Levy, H. Public response to Obamacare on Twitter. *J. Med. Internet Res.* **2017**, *19*, e167. [CrossRef] [PubMed]
38. Deepu, S.; Raj, P.; Rajaraajeswari, S. A Framework for Text Analytics using the Bag of Words (BoW) Model for Prediction. In Proceedings of the 1st International Conference on Innovations in Computing & Networking (ICICN16), Bangalore, India, 12–13 May 2016.
39. Mohammad, S.M.; Kiritchenko, S.; Zhu, X. NRC-Canada: Building the state-of-the-art in sentiment analysis of tweets. *arXiv* **2013**, arXiv:13086242.
40. Ganesh, J.; Gupta, M.; Varma, V. Interpretation of semantic tweet representations. *arXiv* **2017**, arXiv:170400898.
41. Mikolov, T.; Sutskever, I.; Chen, K.; Corrado, G.S.; Dean, J. Distributed representations of words and phrases and their compositionality. In Proceedings of the Advances in Neural Information Processing Systems 26, 27th Annual Conference on Neural Information Processing Systems 2013, Lake Tahoe, NV, USA, 5–10 December 2013.
42. Hill, F.; Cho, K.; Korhonen, A. Learning distributed representations of sentences from unlabelled data. *arXiv* **2016**, arXiv:160203483.
43. Le, Q.; Mikolov, T. Distributed representations of sentences and documents. In Proceedings of the 31st International Conference on Machine Learning (ICML-14), Bejing, China, 22–24 June 2014.
44. Wong, C.A.; Polsky, D.E.; Jones, A.T.; Weiner, J.; Town, R.J.; Baker, T. For third enrollment period, marketplaces expand decision support tools to assist consumers. *Health Aff.* **2016**, *35*, 680–687. [CrossRef]

© 2019 by the authors. Licensee MDPI, Basel, Switzerland. This article is an open access article distributed under the terms and conditions of the Creative Commons Attribution (CC BY) license (http://creativecommons.org/licenses/by/4.0/).

Article

Gender Classification Using Sentiment Analysis and Deep Learning in a Health Web Forum

Sunghee Park and Jiyoung Woo *

Department of Future Convergence Technology, Soonchunhyang University, Asan-si 31538, Korea; sunghee@sch.ac.kr
* Correspondence: jywoo@sch.ac.kr

Received: 15 February 2019; Accepted: 19 March 2019; Published: 25 March 2019

Featured Application: This work can be applied to detect the gender of online users who do not disclose this information.

Abstract: Sentiment analysis is the most common text classification tool that analyzes incoming messages and tells whether the underlying sentiment is positive, negative, or neutral. We can use this technique to understand people by gender, especially people who are suffering from a sensitive disease. People use health-related web forums to easily access health information written by and for non-experts and also to get comfort from people who are in a similar situation. The government operates medical web forums to provide medical information, manage patients' needs and feelings, and boost information-sharing among patients. If we can classify people's emotional or information needs by gender, age, or location, it is possible to establish a detailed health policy specialized into patient segments. However, people with sensitive illness such as AIDS tend to hide their information. Especially, in the case of sexually transmitted AIDS, we can detect problems and needs according to gender. In this work, we present a gender detection model using sentiment analysis and machine learning including deep learning. Through the experiment, we found that sentiment features generate low accuracy. However, senti-words give better results with SVM. Overall, traditional machine learning algorithms have a high misclassification rate for the female category. The deep learning algorithm overcomes this drawback with over 90% accuracy.

Keywords: sentiment analysis; gender classification; machine learning; deep learning; medical web forum

1. Introduction

Sentiment analysis is contextual mining of text that identifies and extracts subjective information in source material, helping a business to understand the social sentiment of their brand, product, or service while monitoring the online conversation. With recent advances in machine learning, text mining techniques have improved considerably. Creative use of advanced artificial intelligence techniques can be an effective tool for doing in-depth research.

Sentiment analysis of web content is becoming increasingly important due to augmented communication through Internet sources such as e-mail, websites, forums, and chat rooms. By collecting these articles and analyzing people's emotions expressed in them, we can figure out people's feelings and opinions about polices, products, brands, and so on. Compared to traditional surveys and other research techniques, this rich information can be obtained with less cost and effort. People can search for particular information based on their individual needs. Patients, patients' significant others, or caregivers also use health-related web forums to get health and medical information and to get comfort from people who are similar to themselves. They also ask questions about the disease and find information that is easy to understand [1]. The medical forum also reflects their feelings [2]. The government relies on web forums to act as a helpdesk to promote people's well-being. Some governments directly operate medical web forums to provide medical information, get to know patients' needs, manage their emotions, and help people share information [3]. Users divulge some pieces of personal information when they sign up, and this information can be used to establish national health care policies. However, for sensitive disease such as AIDS, patients tend not to expose their personal information. In terms of health policy, it is important to know who is looking for more information. If we can guess the information from the non-disclosed data, we can understand who is suffering from a specific disease and what types of worries they might have. Especially, in sensitive diseases, by understanding patients we can help them deal with their situation properly and adapt to society. The analysis of AIDS patients' communication could be useful from a government perspective, too.

Categorizing and analyzing according to demographic information such as gender, age, and region is essential to obtain information such as consumers' emotions, values, and attitudes in all areas of marketing. In particular, it is important to distinguish gender for detailed policy establishment because the situation and necessity can be different by gender. It can also help users to see what topics are most talked about by males and females, and what services are liked or disliked by men and women [4]. Knowing this information is crucial for market intelligence because the information can be used in targeted advertising and service improvement.

Our study is based on a real-life application of a web forum. In this study, we collect data from the AIDS-related bulletin board at Healthboards.com [5], which is one of the top 20 health websites according to Consumer Reports Health Web Watch. Under the premise that "Emotions expressed by men and women will be different," we propose a model that extracts words and emotions from text messages and distinguishes the gender. By establishing a learning model using gender information, we figure out unclassified gender information and identify the gender awareness of AIDS patients by gender.

2. Related Works

In most previous studies on gender classification, the various features fed to machine learning algorithms. At the Islamic Women's political forum, Zhang et al. built a machine learning model that classifies genders by properly combining the characteristics of vocabulary, syntax, structure, uni-gram, and bi-grams [6]. In the study of Ryu et al. [7], they used logistic regression and SVM (a support vector machine) to estimate the gender, age, and location of Twitter users. Wang et al. [8] suggested research based on the Latent Dirichlet Allocation (LDA) model using text data from Facebook or Twitter, in which women have been shown to use a lot of personal themes while men tend to post a lot of philosophical or informative text. In the study of Na and Cho [9], they surveyed the emotions of male and female college students and showed that gender can be distinguished by analyzing emotions using Fastcluster.

In the study of Yan et al. [10], they presented a naïve Bayes classification approach to identify the gender of weblog authors. They used weblog-specific features such as webpage background colors and emoticons. They report an F-measure of around 64% using their features. Mukherjee and Liu. [11] proposed two new techniques. They used their own POS Sequence mining algorithm and an Ensemble feature selection technique, and achieved an accuracy of 88%.

Pennacchiotti and Popescu [12] proposed a user classification model applying machine learning algorithm to the feature set including user profile, user tweeting behavior, linguistic content of user messages, and user social network features. They explored the feature importance and found that linguistic features are consistently effective in user classification.

Dwivedi et al. [13] present two systems, a manual feature extraction system and a deep learning method, to automatically classify the gender of blog authors. For the deep-learning-based model, they apply a Bidirectional Long Short-Term Memory Network. Barlte and Zheng [14] report an accuracy of 86% in gender classification on blog datasets by applying deep learning models based on the Windowed Recurrent Convolutional Neural Network (WRCNN).

Filho et al. [15] proposed textual meta-attributes, taking into account the characters, syntax, words, structure, and morphology of short, multi-genre, content-free text posted to Twitter to classify an author's gender via three different machine learning algorithms. The novel contribution of this work is to employ a word-based meta-attribute such as Ratio between hapax dislegomena (a word that appears only twice in a whole text) and the total number of words. Furthermore, they developed a textual morphology based on the meta-attributes from textual structures such as the ratio between the number of pronouns and the total number of words. This work achieved 81% accuracy, but the performance by each class is not presented. Garibo-Orts [16] proposed a statistical approach to the task of gender classification in tweets. The statistical features include skewness, kurtosis, and central moments; these statistical features make the learning algorithm context-free and language-independent.

Recently, Convolutional Neural Networks (CNNs) have also been successful in various text classification tasks. Kim [17] showed that a simple CNN with little hyper-parameter tuning and static vectors achieves excellent results. Severyn et al. [18] proposed a deep learning model for Twitter sentiment classification. They advanced a CNN model to adjust word embedding using unsupervised learning.

Based on previous works, females and males are found to have differences in terms of the vocabulary used and the emotional expression. During a literature review, we found some limitations in the current literature. First, most recent works focus on Twitter or weblogs. Even though the gender information disclosed in the medical web forum, especially on sexually transmitted diseases, is important, we hardly found any work dealing with medical web forums. Secondly, studies using sentiment features derived from text are lacking, while word features are explored a lot. Thirdly, deep learning is applied a lot to text classification, but its application to gender classification from textual information is rare.

In this work, we aim to figure out how the sentiment features derived from emotions expressed in articles work in gender classification. We will do this by using both machine learning and deep learning methods.

3. Experiment

We retrieved the users' gender information from an AIDS-related medical web forum and presented a gender detection model that classifies gender based on the emotions expressed in posts and comments. We developed the sentiment feature set that expresses how often the posts contain emotions, and assessed the emotional complexity, which is the number of emotion categories shown in a post. Using vocabulary characteristics and emotions from the disclosed data, we built machine learning models and measured the accuracy, varying the feature sets to select the best model. The proposed framework is presented in Figure 1.

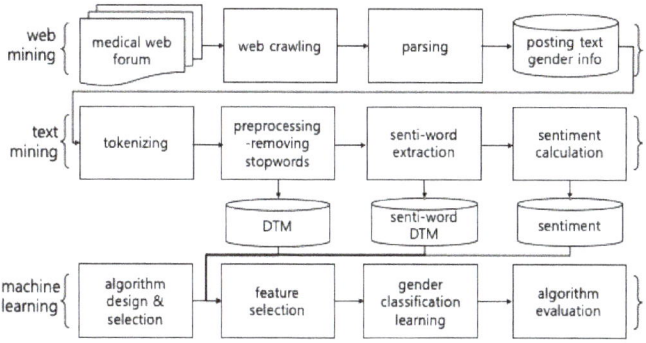

Figure 1. Gender classification model for medical web forum.

3.1. Dataset

From 14 November 2000 to 25 February 2010, we collected 3451 posts and 18,944 comments—in total, 22,395 messages—from an AIDS-related bulletin board from HealthBoard.com. It is one of the first independent health community websites and has more than 280 bulletin boards for communication about diseases. As shown in Table 1, this dataset has 8954 male-written messages and 6078 female-written messages. A total of 3282 users participated in the AIDS bulletin board, with 849 female users and 1054 male users. This implies that 1379 users have not disclosed their gender. The average post length is 113 words for men and 106 words for women. The percentage of users who disclosed gender information is 59% for males and 41% for females; 33% of the total has no gender information. As for the Alzheimer's-related board, only 1708 (6.2%) out of 27,874 (posts: 2486, comments: 25,134) have not disclosed their gender. The number of users in the Alzheimer's bulletin board is 1498. The number of female users is 924 and male users 185. In total, 1109 users have gender information. Compared to the users in the Alzheimer's-related board, users in the AIDS-related board have a low ratio of disclosed gender information.

Table 1. The gender information disclosed in an AIDS-related web forum.

Gender	Numbers of Messages	Ratio (%)
Male	8954	40
Female	6078	27
N/A	7363	33
Total	22,395	100

3.2. Feature Extractions

Before feature extraction, we first perform data preprocessing. We first converted everything to lowercase, removing punctuations, numbers, and stopwords. Stopwords are words that have no special meaning but are frequently used. For example, in English, "a," "the," and "is" are examples of stopwords. In this work, we excluded these words to reduce the feature set in text mining, with the assumption that the stopwords are no different in distribution between women and men.

Then, we build the tokenizer only on the training data. We could add the testing data, but results may go up or down. Testing is needed to determine whether or not adding the testing data will help. In most text classification studies, words are the most important features. Here are the features that we considered in this work.

3.2.1. Unigram

We extracted the word features from the text to construct a learning model. The first step was to derive words after preprocessing. We used the Bag-of-Words technique to tokenize each word into a DocumentTermMatrix that indicates how many times the word appears in each article. The Bag-of-Words technique transforms the sentence into a set of numeric vectors as follows. First, the word set is built by collecting words from all sentences in all documents. Then, a sentence is expressed as the word counts in word order. For example, in "symptoms appear after at least two" and "it was two years of unnecessary stress," these two sentences generate the word set, {symptoms, appear, after, at, least, two, it, was, years, of, unnecessary, stress}. Two sentences are expressed as the counts of each word as {1,1,1,1,1,1,0,0,0,0,0,0} and {0,0,0,0,0,1,1,1,1,1,1,1}, respectively. The numeric vector is derived for a document by aggregating the word count for all sentences in a document. The document term matrix (DTM) is represented in Figure 2.

		Terms							
		Condom	Get	Give	Good	Help	Hiv	however	impossible	...
Documents	Post1	1	2	1	1	1	1	1	1	
	Post2	0	1	0	0	1	1	0	0	
	Post3	0	1	0	0	2	11	0	0	
	Post4	0	5	0	2	1	3	0	0	
	...									

Figure 2. Representation of DocumentTermMatrix.

We extracted the document term matrix in two ways, the word occurrence and the occurrence frequency, to test which feature extraction method is efficient.

To reduce the feature set, we set the threshold value to the number of appearances for extracting word features. We incrementally increased the threshold value from 5 to 15 by 5s and checked the classification performance.

3.2.2. Sentiment Features

To derive the sentiment features, we used tidytext, which is a dictionary package built in R, to measure the rate of emotions. There are several dictionaries, but we used NRC [19] and BING. The NRC dictionary has 10 categories: trust, fear, negative, sadness, anger, surprise, positive, disgust, joy, and anticipation. Based on this, we developed sentiment-related features: the number of emotion types expressed in a message, objectivity (1-number of used emotion types/total emotion (10)), and emotional complexity (the number of emotion types/10).

The Bing dictionary classifies words only into positive or negative, and has 6788 words. Based on the Bing dictionary, we calculated the positive rate (the number of positive words/the total number of words), negative rate (the number of negative words/the total number of words that contain emotions), and the total number of words.

As shown in Tables 2 and 3, women use the words 'shine,' 'thank,' 'bless,' and 'glad' (positive words) and 'problem,' 'scary,' and 'illness' (negative words) about twice as often as men. On the other hand, men use the words 'accurate,' 'important,' 'receptive' (positive words) and 'issue,' 'fever,' and 'aches' (negative words) more than twice as often as woman.

Table 2. Differences in positive words between men and women.

Word	Men	Women
Shine	1	419
Thank	655	714
Glad	154	233
Bless	168	267
Accurate	350	166
Important	196	76
Correct	149	54

Table 3. Differences in negative words between men and women.

Word	Men	Women
Problem	246	337
Scary	80	147
Illness	144	211
Issue	565	179
Fever	358	140
Aches	184	39

3.3. Classification Algorithms

We employed a representative machine learning algorithm for text classification.

3.3.1. Naïve Bayes

Naïve Bayes is the first approach we tried. Let $C = (c_1, c_2)$ be the gender class, and $F = (f_1, f_2, \ldots f_n)$ are features, according to Bayes's theorem:

$$P(c|F) = \frac{P(c)P(F|c)}{P(F)}. \tag{1}$$

The naïve Bayes assumption is that:

$$\hat{P}(F|c) = \prod_{i=1}^{n} \hat{P}(f_i|c). \tag{2}$$

Based on the naïve Bayes assumption, the probability of belonging to each class is as follows:

$$\hat{P}(F|c) = \operatorname*{argmax}_{c} P(C = c|F) = \operatorname*{argmax}_{c} P(C|c) \prod_{i=1}^{n} P(f = f_i|C = c). \tag{3}$$

The basic assumption is that all features are independent. This assumption is too strong, but it works well in reality.

3.3.2. Support Vector Machine

The support vector machine is known as a powerful algorithm until deep learning algorithm replaced it [20]. Various lines can be drawn between classes A and B. If new data comes in and overlaps with a line that is on one side, it is hard to judge whether those data belong to class A or class B. Therefore, the best way to increase the classification accuracy is to deploy a line in a center between groups with the maximum distance from the line to groups. The maximum distance from the line to groups is called the margin. At the maximum margin, the middle boundary is called the optical hyper-plane.

3.3.3. Random Forest

Random forest is an algorithm that generates multiple decision trees by randomly sampled data for learning and assembles the results of the decision trees by majority voting. It is a type of ensemble leaning method that implements classification, regression, and clustering based on group learning. Random forest trains the model using the subset of features and the subset of data in a repetitive way. This operation reduces the risk of overfitting. This algorithm has been widely used in various applications in recent years and has been proved to outperform other algorithms.

3.4. Deep Learning Approach

The deep learning algorithm is a kind of machine learning. Machine learning focuses on solving real-world problems with neural networks created by mimicking the decision-making processes that take place in the human brain by adopting some of the main ideas of artificial intelligence. Deep learning focuses more on the tools and techniques of machine learning. The difference between machine learning and deep learning is that, in the course of machine learning, people are still involved in the process of extracting features from the given data to train the machine, but deep learning uses the given data as input data. It does not feature an engineering process to extract the features of training by human intervention, but automatically performs learning of important features in the data itself. So deep learning is called end-to-end machine learning. This is because the machine learns from beginning to end, by itself, without human intervention.

3.4.1. Convolution Neural Network

CNNs are widely used on image datasets [21,22], but researchers have found that they work great with text, too. Previous works showed that CNN outperforms the state-of-art algorithm in many cases [21,22] That is because text can be considered a 1D image. Now, the explanation for why CNNs work for images is quite complicated, but they (partly) revolve around something called 'spatial locality'. What this means is that CNNs can find patterns not just from each feature (e.g., a pixel) but from locations of pixels (e.g., neighborhoods of pixels correlated with another). This makes sense because images are not a random collection of pixels, but pixels are connected across the image.

3.4.2. Convolution Neural Network (CNN) for Text Classification

To make the sentences in text into an image, we need to encode words with numeric values. Word embedding is employed to express words as numeric values. The method that assigns a unique integer value to each word would cause the dimension explosion. Word embedding set the dimensionality, k, and expresses the word using k-dimensional values. For example, when k is set to 5, "hospital" is expressed as {0.1,0.2,0.3,0.8,0}. This embedding method dramatically reduces the dimensions. However, this method does not consider the relationship between words when placing a word in a matrix. The advanced technique is to consider the relationship between words when transforming the input data into an input matrix like an image. Word2Vec considers that words that appear in a similar context have similarities in semantic meaning. We adopted Google's word2Vec [23], which is a pretrained model from the GoogleNews dataset.

We performed the experiment using word-embedding method and Word2Vec method as well.

The architecture of the text classification is shown in Figure 3. We transformed words into a k-dimensional vector. We set the number of words to make an input matrix, denoted as n. A sentence is tokenized and n words organize the input data. At the end of the sentence, there could not be enough words to fill the input. In this case, we pad the rows with zeros until the number n is filled.

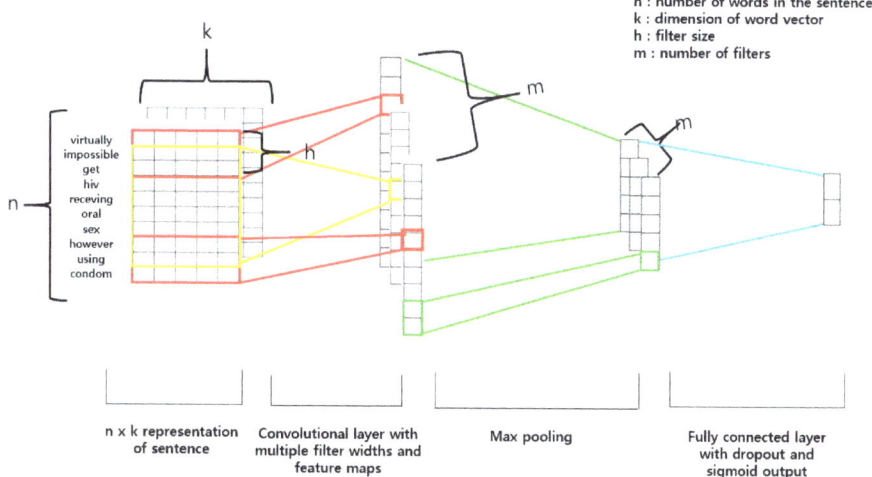

Figure 3. Model architecture with an example sentence.

To derive abstract features from input data and make the algorithm derive useful features automatically, we employ a hidden layer and a convolutional layer to the network.

In a convolution layer, we feed the input matrix to the filter of k*h size, then the input of h*k size is mapped to a cell in the convolutional layer. In Figure 3, the rectangle with a red line is a 3*k filter. The filter works as an aggregator of h words. The filter needs to keep the word meaning when aggregating words, so the horizontal size of the filter is set to k. As a result, the convolutional layer is one-dimensional. The number of nodes in a convolution layer is calculated as n-h + 1.

The vertical size of the filter can be varied. In this work, we test two sizes of filter, 3 (in red) and 8 (in yellow). The number of filters depends on the user's decision. More filters are needed to enrich features. The max pooling layer is also employed to avoid overfitting a result of too much features. The max pooling layer adopts a filter with the size of (n-h + 1)*1 and finds the maximum value for all values in a precedent convolutional layer. The max pooling layer generates one-dimensional m rows. Finally, m rows are fully connected to the output. The output layer has two nodes to express bi-class, gender in our case. Two nodes are filled with 1 or 0, it indicates male or female respectively.

4. Results

4.1. Traditional Machine Learning Results

Initially, 70% of data, including gender information, were used as training data and 30% as test data. We experimented with naïve Bayes (NB), support vector machine (SVM), and random forest (RF), which are representative algorithms proven to outperform other algorithms. The precision measure and the recall measure are calculated for each class. For the female class, the precision is calculated as TP/(TP + FP) and recall is calculated as TP/(TP + FN). For the male class, the precision and recall are calculated as FN/(FN + TN) and TN/(FP + TN), respectively. The positive class is defined as the class of interest. In this work, the female class and male class are treated equally, but the positive class is set as the female class. The accuracy, the true positive rate, and the true negative rate are calculated considering two classes. Accuracy indicates how well a binary classification test correctly identifies each class. False positive rate is calculated as FP/(FP + TN) and indicates the ratio of the wrongly classified females among the males. False negative rate is calculated as FN/(FN + TP). Using only sentiment features, overall, the female class has poor performance. Three algorithms did not recognize the female classes well and classified most of the users into the male class. This indicates that sentiment features such as emotion types and their density are not effective for classifying authors' gender on their own.

To test the effects of sentiment features, we performed three experiments as follows. At first, we only built a sentiment feature set including the ratio of 10 types of emotional occurrences, objectivity, and emotional complexity. The accuracy of them are 58.33% (NB), 60.86% (SVM), and 58.66% (RF), respectively, as shown in Table 4. Compared to the male class, the female class has a poor performance. Over 50% of female users were classified into the male class.

Table 4. The accuracy of gender detection using sentiment feature set.

Real \ Algorithm	NB		SVM		RF	
	Female	Male	Female	Male	Female	Male
Female	174 (True positive: TP)	200 (False negative: FN)	81	42	402	413
Male	1679 (False positive: FP)	2456 (True negative: TN)	1723	2663	1451	2243
Precision	9.39%	92.47%	4.49%	98.45%	21.69%	84.45%
Recall	46.52%	59.40%	65.85%	60.72%	49.33%	60.72%
Accuracy		58.33%		60.86%		58.66%
False positive rate		40.60%		39.28%		39.28%
False negative rate		53.48%		34.15%		50.67%

The second experiment was performed using senti-word features. Instead of calculating sentiment value from senti-word, we checked the senti-word occurrence. Among the 7954 words that appeared more than five times each, we selected only 1594 senti-words as features that are in the NRC dictionary. The accuracy increased a little bit with the SVM and RF algorithms, as shown in Table 5. The precision for the female class improved. The senti-words used to express the author's emotion were comparatively effective for gender classification. Among the three algorithms, SVM had the highest performance for the sentiment feature set and the senti-word feature set.

Table 5. The accuracy of gender detection using senti-word feature set.

	NB		SVM		RF	
	Female	Male	Female	Male	Female	Male
Female	1706	2434	701	236	514	210
Male	117	252	1122	2450	1309	2470
Precision	93.58%	9.38%	38.45%	91.21%	28.20%	92.16%
Recall	41.21%	68.29%	74.81%	68.59%	70.99%	65.36%
Accuracy	43.42%		69.88%		66.27%	
False positive rate	31.71%		31.41%		34.64%	
False negative rate	58.79%		25.19%		29.01%	

The final experiment was performed using the entire feature set, combining word features and sentiment features. When we assessed all 7954 words that occurred more than five times along with the sentiment features, the accuracy improved. However, the performance of SVM rather decreased. In this case, RF has the highest performance and the precision for the female class somewhat improved. However, the female class still has a poor performance as shown in Table 6.

Table 6. The accuracy of gender detection using entire feature set (word features + sentiment features).

	NB		SVM		RF	
	Female	Male	Female	Male	Female	Male
Female	1838	2623	1	0	754	135
Male	14	34	1822	2686	1069	2551
Precision	99.24%	1.28%	0.05%	100.00%	41.36%	94.97%
Recall	41.20%	70.83%	100.00%	59.58%	84.81%	70.47%
Accuracy	41.52%		59.55%		73.33%	
False positive rate	29.17%		40.42%		29.53%	
False negative rate	58.80%		0.00%		15.19%	

Based on the three experiments varying the feature set, we obtained two findings, one is the algorithm perspective and the other is the feature perspective. We can conclude that SVM underperforms other algorithms in the case of a few features. On the other hand, RF works well with more features. In addition, we can conclude that NB is not appropriate for text classification task. For feature perspective, words are good features for gender classification in the web forum. Types of emotion and their density are not enough, although these sentiment features reduce the computational complexity with a small number of features. However, when using with SVM, the senti-word feature set generates the comparable performance to the word feature set.

Even though words are good features, we cannot employ all words because we need to consider the computational complexity and cost to make the algorithm work. Machine learning algorithms support a limited number of features, unlike the deep learning algorithm, and we need to reduce the number of words included in the feature set. For feature reduction, we explored the threshold value to extract words from 5 to 100, as shown in Table 7. The number of words involved in learning is displayed at the end of Table 7. We also tested which feature extraction method among the occurrence with boolean and the term frequency is useful. From the experiments results, the occurrence with boolean outperforms the term frequency. In naïve Bayes, which assumes the independence of the variables, the number of words decreases when setting a higher threshold value, and the performance improves. SVM has the opposite behavior with NB. The best performance is achieved when applying the RF model to the term frequency features with a threshold value of 10, resulting in 4822 words. However, the performance of RF begins to decline beyond a certain number of features from the first two rows of Table 7. From all the experiments, we could conclude that traditional machine learning algorithms are not good at gender classification, resulting in a large performance gap between two classes. In the next section, we will see how much the deep learning algorithm improved the performance.

Table 7. The accuracy of gender detection using different feature set.

Feature | Accuracy	NB (%)	SVM (%)	RF (%)
TF/Boolean (5) [1] + senti	41.52/41.32	59.55/59.57	73.3/72.79
TF/Boolean (10) [2] + senti	44.95/45.13	70.68/71.66	**74.41/73.59**
TF/Boolean (15) [3] + senti	49.48/51.08	71.72/**72.79**	73.16/73.05
TF/Boolean (100) [4] + senti	**61.5/66.16**	71.99/72.52	72.9/73.52

[1] including 7588 words, [2] including 4822 words, [3] including 4732 words, [4] including 1108 words.

4.2. Deep Learning Results

We tested two methods of embedding words into vectors, random word-embedding and the word2Vec method. We also varied the CNN structure with two filter sizes of length 3 or 8, which implies that three words or eight words are aggregated. The number of filters is set to 10. For parameter optimization, we used ADAM [24]. This uses squared gradients to scale the learning rate and takes advantage of momentum by using the moving average of the gradient instead of the gradient itself. The experiment results show that the random word-embedding method outperforms the word2Vec method.

We guess that word2Vec is built with formal text from GoogleNews, but our text dataset is written in spoken language. Thus, word2Vec does not work well in our case. The accuracy of the random word embedding method ranges between 88% and 91% depending on the CNN structure. On the other hand, the accuracy with the word2Vec ranges between 67% and 71%.

Secondly, we varied the feature set from word features to senti-word features. The sentiment feature set is low-dimensional with a small number of features, so we judged that deep learning is not necessary for this feature set. Thus, we tested CNN for the reduced feature consisting of senti-words and the full feature set consisting of word features. We achieved 88.7% with senti-word features and 90.6% with word features. The number of senti-word features is at 20% of the number of word features, but it generates a comparable performance. Compared to traditional machine learning algorithms, CNN dramatically increases the performance. For the female class, the precision and recall are much improved and the performance for the two classes is equally good.

We changed the CNN structure to find the best structure for our context. Starting from a simple structure, we designed more complicated structures. The combination of two convolution layers proceeded by the pooling layer, the dropout layer after that, and the fully-connected layer at last, outperformed other structures. A more complicated model with two convolutional layers outperforms the simplest model with one convolutional layer. However, the accuracy increases by at most 0.004. Slight modifications to the baseline model, in the second row, degrade the accuracy. From the other models, we found one fully connected layer is enough for comparing the second through fourth models. The max pooling layer employed to reduce feature dimension is best deployed right after each convolutional layer. In future work, we will test more complicated structures and more diversity in terms of the number of filters and sizes. Table 8 indicates the result of CNN performance according to each structure.

Table 8. Comparison of CNN performance according to the structures.

CNN Structure	Accuracy	Female		Male	
		Precision	Recall	Precision	Recall
Conv + Pool + Dropout + FC	0.906	0.93	0.91	0.87	0.90
Conv + Pool + Conv + Pool + Dropout + FC	**0.910**	0.93	0.92	0.88	0.89
Conv + Pool + Conv + FC + Dropout + FC	0.905	0.94	0.89	0.86	0.92
Conv + Conv + Pool + FC + Dropout + FC	0.887	0.91	0.90	0.85	0.87

5. Discussion and Conclusions

In this study, we constructed a model to detect gender information by using machine learning algorithms based on word and sentiment feature sets. We developed various sentiment feature sets including the number of emotion types expressed in a message, objectivity, and emotional complexity.

To check the effectiveness of sentiment features in gender classification, we varied the feature sets, including sentiment, senti-word, sentiment+word feature sets and using machine learning algorithms or deep learning algorithms. From the experiments, we found that sentiment features with a small number of features and domain-independent are not enough for classifying gender in the medical web forum. Words should be incorporated to classify the author's gender. We found that it is possible to construct gender classification models by using a sentiment+word feature set. In cases of a lightweight system with a limited number of features, SVM performed moderately with senti-word features.

However, overall, the traditional machine learning algorithm failed, having low precision in the female class. Many male authors were misidentified as female authors. When we applied the deep learning algorithm, both gender classes were identified well. The senti-word feature set and word feature set both worked well, with a 2% accuracy gap. The number of senti-word features is 20% of the number of word features, but generates a comparable performance.

As a design issue, to embed text into a matrix form, we tested word-embedding methods, one random and the other pre-built word2Vec. To find the better model, we varied the CNN structure by adding and deleting hidden layers, the convolutional layer, and the max pooling layer.

This work poses remaining challenges as follows. We tested the word2Vec built by Google and found that it does not work well for our dataset. In future work, we will build our own word2Vec suitable for medical web forums by adopting Google's training method. A consideration of word similarity would improve the detection performance. Secondly, in the current work, we eliminated stopwords, numbers, and special characters to reduce the feature set. However, previous work has suggested that these words could be helpful in the text classification task. In the deep learning model, feature reduction is not necessary, so we will check their effectiveness. In addition, we intend to add features in the context of social networks by analyzing the relationship between the posts and comments in order to improve the accuracy of the proposed model. The social network features are also generic across disease types and forums.

Author Contributions: S.P. performed the experiments and writing. J.W. collected the dataset and participated in writing.

Funding: This work was supported by the research fund of Soonchunhyang University (project number 20171094) and a National Research Foundation of Korea (NRF) grant funded by the Korean government (MSIP; Ministry of Science, ICT & Future Planning (NRF-2017R1D1A3B03036050)).

Acknowledgments: The authors gratefully acknowledge the support by the research fund of Soonchunhyang University (project number 20171094) and a National Research Foundation of Korea (NRF) grant funded by the Korean government (MSIP; Ministry of Science, ICT & Future Planning (NRF-2017R1D1A3B03036050)).

Conflicts of Interest: The authors have no conflicts of interests.

References

1. Weaver, J.B., III; Mays, D.; Weaver, S.S.; Hopkins, G.L.; Eroğlu, D.; Bernhardt, J.M. Health information–seeking behaviors, health indicators, and health risks. *Am. J. Public health* **2010**, *100*, 1520–1525. [CrossRef] [PubMed]
2. Woo, J.; Lee, M.J.; Ku, Y.; Chen, H. Modeling the dynamics of medical information through web forums in medical industry. *Technol. Forecast. Soc. Chang.* **2015**, *97*, 77–90. [CrossRef]
3. Denecke, K.; Nejdl, W. How valuable is medical social media data? Content analysis of the medical web. *Inf. Sci.* **2009**, *179*, 1870–1880. [CrossRef]
4. Sullivan, C.F. Gendered cybersupport: A thematic analysis of two online cancer support groups. *J. Health Psychol.* **2003**, *8*, 83–104. [CrossRef] [PubMed]
5. Healthboard. Available online: https://www.healthboards.com/ (accessed on 25 March 2019).

6. Zhang, Y.; Dang, Y.; Chen, H. Gender classification for web forums. *IEEE Trans. Syst. Man Cybern. Part A Syst. Hum.* **2011**, *41*, 668–677. [CrossRef]
7. Ryu, K.; Jeong, J.; Moon, S. Inferring Sex, Age, Location of Twitter Users. *J. KIISE* **2014**, *32*, 46–53.
8. Wang, Y.-C.; Burke, M.; Kraut, R.E. Gender, topic, and audience response: An analysis of user-generated content on facebook. In Proceedings of the SIGCHI Conference on Human Factors in Computing Systems, Paris, France, 27 April–2 May 2013; pp. 31–34.
9. Na, Y.; Cho, G. Grouping preferred sensations of college students using sementic differential methods of sensation words. *Korean J. Sci. Emot. Sensib.* **2002**, *5*, 9–16.
10. Yan, X.; Yan, L. Gender Classification of Weblog Authors. In Proceedings of the AAAI Spring Symposium: Computational Approaches to Analyzing Weblogs, Palo Alto, CA, USA, 27–29 March 2006; pp. 228–230.
11. Mukherjee, A.; Liu, B. Improving gender classification of blog authors. In Proceedings of the 2010 Conference on Empirical Methods in Natural Language Processing, Cambridge, MA, USA, 9–11 October 2010; pp. 207–217.
12. Pennacchiotti, M.; Popescu, A.-M. A machine learning approach to twitter user classification. In Proceedings of the Fifth International AAAI Conference on Weblogs and Social Media, Barcelona, Spain, 17–21 July 2011.
13. Dwivedi, V.P.; Singh, D.K.; Jha, S. Gender Classification of Blog Authors: With Feature Engineering and Deep Learning using LSTM Networks. In Proceedings of the 2017 Ninth International Conference on Advanced Computing (ICoAC), Chennai, India, 14–16 December 2017; pp. 142–148.
14. Bartle, A.; Zheng, J. *Gender Classification with Deep Learning*; Stanford cs224d Course Project Report; The Stanford NLP Group: Stanford, CA, USA, 2015.
15. Lopes Filho, J.A.B.; Pasti, R.; de Castro, L.N. Gender classification of twitter data based on textual meta-attributes extraction. In *New Advances in Information Systems and Technologies*; Springer: Berlin/Heidelberg, Germany, 2016; pp. 1025–1034.
16. Garibo-Orts, O. A big data approach to gender classification in twitter. In Proceedings of the Ninth International Conference of the CLEF Association (CLEF 2018), Avignon, France, 10–14 September 2018.
17. Kim, Y. Convolutional neural networks for sentence classification. *arXiv* **2014**, arXiv:1408.5882.
18. Severyn, A.; Moschitti, A. Unitn: Training deep convolutional neural network for twitter sentiment classification. In Proceedings of the 9th International Workshop on Semantic Evaluation (SemEval 2015), Denver, CO, USA, 4–5 June 2015; pp. 464–469.
19. Mohammad, S.M. Challenges in sentiment analysis. In *A Practical Guide to Sentiment Analysis*; Springer: Berlin/Heidelberg, Germany, 2017; pp. 61–83.
20. Nayak, J.; Naik, B.; Behera, H. A comprehensive survey on support vector machine in data mining tasks: Applications & challenges. *Int. J. Database Theory Appl.* **2015**, *8*, 169–186.
21. Zhang, Y.D.; Dong, Z.; Chen, X.; Jia, W.; Du, S.; Muhammad, K.; Wang, S.H. Image based fruit category classification by 13-layer deep convolutional neural network and data augmentation. *Multimedia Tools Appl.* **2019**, *78*, 3613. [CrossRef]
22. Wang, S.H.; Sun, J.; Phillips, P.; Zhao, G.; Zhang, Y.D. Polarimetric synthetic aperture radar image segmentation by convolutional neural network using graphical processing units. *J. Real-Time Image Process.* **2018**, *15*, 631. [CrossRef]
23. word2Vec. Available online: https://code.google.com/archive/p/word2vec/ (accessed on 25 March 2019).
24. Kingma, D.P.; Ba, J.L. Adam: A method for stochastic optimization. *arXiv* **2014**, arXiv:1412.6980v9.

© 2019 by the authors. Licensee MDPI, Basel, Switzerland. This article is an open access article distributed under the terms and conditions of the Creative Commons Attribution (CC BY) license (http://creativecommons.org/licenses/by/4.0/).

Article

Personality or Value: A Comparative Study of Psychographic Segmentation Based on an Online Review Enhanced Recommender System

Hui Liu [1,†], Yinghui Huang [1,2,†], Zichao Wang [3], Kai Liu [1], Xiangen Hu [1,4] and Weijun Wang [1,*]

1. Key Laboratory of Adolescent Cyberpsychology and Behavior, Ministry of Education, Central China Normal University, Wuhan 430079, China; huiliu931031@gmail.com (H.L.); yinghui0121@mails.ccnu.edu.cn (Y.H.); ccnulk@mail.ccun.edu.cn (K.L.); xiangenhu@gmail.com (X.H.)
2. School of Information Management, Central China Normal University, Wuhan 430079, China
3. Department of Electrical and Computer Engineering, Rice University, Houston, TX 77005, USA; wangzichao6@gmail.com
4. Department of Psychology, The University of Memphis, Memphis, TN 38152, USA
* Correspondence: wangwj@mail.ccnu.edu.cn; Tel.: +86-153-0715-0076
† Hui Liu and Yinghui Huang are co-first authors.

Received: 1 March 2019; Accepted: 8 May 2019; Published: 15 May 2019

Abstract: Big consumer data promises to be a game changer in applied and empirical marketing research. However, investigations of how big data helps inform consumers' psychological aspects have, thus far, only received scant attention. Psychographics has been shown to be a valuable market segmentation path in understanding consumer preferences. Although in the context of e-commerce, as a component of psychographic segmentation, personality has been proven to be effective for prediction of e-commerce user preferences, it still remains unclear whether psychographic segmentation is practically influential in understanding user preferences across different product categories. To the best of our knowledge, we provide the first quantitative demonstration of the promising effect and relative importance of psychographic segmentation in predicting users' online purchasing preferences across different product categories in e-commerce by using a data-driven approach. We first construct two online psychographic lexicons that include the Big Five Factor (BFF) personality traits and Schwartz Value Survey (SVS) using natural language processing (NLP) methods that are based on behavior measurements of users' word use. We then incorporate the lexicons in a deep neural network (DNN)-based recommender system to predict users' online purchasing preferences considering the new progress in segmentation-based user preference prediction methods. Overall, segmenting consumers into heterogeneous groups surprisingly does not demonstrate a significant improvement in understanding consumer preferences. Psychographic variables (both BFF and SVS) significantly improve the explanatory power of e-consumer preferences, whereas the improvement in prediction power is not significant. The SVS tends to outperform BFF segmentation, except for some product categories. Additionally, the DNN significantly outperforms previous methods. An e-commerce-oriented SVS measurement and segmentation approach that integrates both BFF and the SVS is recommended. The strong empirical evidence provides both practical guidance for e-commerce product development, marketing and recommendations, and a methodological reference for big data-driven marketing research.

Keywords: psychographic segmentation; user preference prediction; lexicon construction; online review; recommender system; big data-driven marketing

1. Introduction

In big data era, the research paradigms of marketing service have been greatly changed by the enormous marketing data accumulated from the internet, such as on demographics, user behavior (we will henceforth use the word "user", "customer", and "consumer" interchangeably), and social relationships. These fine-grained marketing data are informative, thus providing marketers with extra opportunities to evaluate users' preferences [1], predicting the next product users will buy [2,3], delivering targeted advertisements [4,5], uncovering consumer perceptions of brand [6], and acquiring competitive intelligence [7]. In particular, psychological aspects of user behavior, critical for understanding rather than merely predicting consumer preference, ultimately contribute to intelligent decision-making in marketing. Unfortunately, investigations of how big data helps inform has, thus far, only received scant attention [8].

Studies have demonstrated that psychological variables, such as value and personality, are the main theories and tools for user psychographic segmentation, which is an important determinant of user purchase behaviors and preferences [9,10]. In particular, personality traits, one of the psychographic segmentation components, has been the main consumer psychological and marketing characteristic used in e-commerce. Unfortunately, the predictive and explanatory power of personality traits for online user behavior remains controversial [11]. This controversy motivates the investigation of other types of psychographic segmentation, such as value, in understanding user preferences, which have not been investigated. Furthermore, it is unclear whether the predictive and explanatory power varies between product categories and between different segment-wise preference prediction methods. The main reason is that collecting psychographic segmentation data from e-commerce users is difficult on a large scale, typically requiring consumers to complete lengthy questionnaires.

Psychological variables, such as personality and value, are deeply embedded in the language that people use today [12]. With massive user data in the form of natural language, natural language data provides a clearer picture of people's cognitive and behavioral processes than data collected from traditional and widely used self-report surveys [13–15]. Practically, natural language processing (NLP) techniques can be applied to identifying psychographic variables, such as e-commerce users' online word use, to understand and predict users' purchase behaviors and preferences on a large scale.

E-commerce websites, in particular, have accumulated a large amount of user-generated content (UGC), which provides the basis for observing users' psychographics and predicting user preferences directly. With the rapid development of techniques such as big data, artificial intelligence, and computational linguistics, UGC provides a reliable path for automatically identifying consumer psychographics, including personality and values, based on unstructured data. Inspired by recent advances in big data-driven psycholinguistic research, which indicate the behavioral evidence of online word use related to psychographics [16,17], we base our research on real-world Amazon consumer review and rating data. We propose psychographics-related word use evidence, extract consumers' psychographic characteristics using sentiment analysis methods, and introduce a deep neural network (DNN) to predict and explain user preferences in e-commerce.

We found that, overall, psychographic variables significantly improved the explanatory power of e-consumer preferences across most product categories we studied, whereas the improvement in predictive power was not significant. Specifically, the Schwartz Value Survey (SVS) tended to outperform Big Five Factor (BFF) segmentation in predicting and explaining user preferences, with the exception of a few product categories. However, somewhat surprisingly, dividing e-consumers into heterogeneous groups using a clustering method did not significantly improve the predictive and explanatory power of online consumer preferences compared to handling consumers as a whole. Furthermore, the DNN method that we proposed demonstrated the best performance in understanding e-consumer preferences, and regarding product category, there were more significant differences for psychographic segmentation in predicting, than explaining, e-consumer preferences. Additionally, we recommend an e-commerce-oriented SVS measurement and segmentation approach that integrates both BFF and the SVS.

Our work extends the depth and breadth of psychographic theories through user preference prediction in e-commerce scenarios. Specifically, we found that subdimensions of psychographic variables and product types provide practical references for psychographic measurement development and applications for the specific e-commerce product categories that we studied. By introducing psychographic-related word use behavioral evidence, followed by big data approaches, we have attempted to overcome the difficulties of obtaining e-consumer psychographics on a large scale, and have provided a promising psychographic-based consumer preference prediction method for subsequent research.

2. Literature Review and Research Design

In this section, we review the foundational work on consumer psychographics and trace its relationship with user preferences, particularly in online scenarios. Building on the current development in online behavioral measures of psychographics, we then introduce and propose related methods and techniques in computer science and psycholinguistics. Finally, we organize and outline our research questions and framework.

2.1. Psychographic Segmentation

Market segmentation has had a longstanding history since Smith (1956) [18] first suggested it as a product differentiation strategy to increase competitiveness. It has been widely acknowledged as a fundamental tool in understanding customer behaviors [19,20]. Wedel and Kamakura (2012) defined segmentation as "a set of variables or characteristics used to assign potential customers to homogeneous groups" [21]. Segmentation is critical because a company has limited resources and must focus on how to best identify and serve its customers.

Initially, market segmentation was mainly rooted in personality profiles. The most frequently used scale for measuring general aspects of personality as a way to define homogeneous submarkets is the Edwards Personal Preference Schedule. Generally, however, early studies based on Edwards Personal Preference Schedule were plagued with low and even inconsistent correlations with customer behavior and preferences, and hence failed to satisfy marketers' needs [22]. One of the main reasons was that the studies used standardized personality tests that were originally developed in clinical or academic instead of business contexts [23]. Since then, the Big Five Factor personality traits—that frequently and consistently appeared in most attempts to define the basic human factors, that is, neuroticism, extraversion, agreeableness, conscientiousness and openness—have replaced the Edwards Personal Preference Schedule.

Lifestyle is defined as a set of behaviors that mirrors individual psychological considerations and sociological consequences [24], and is more valid in shaping individual behavior compared to personality traits. The lifestyle construct used in market segmentation is based on research on motivation [23] and personal values [22]. Mitchell (1994) proposed that lifestyle is a mixture of a personal life and perceived value, whereas value is a synthesis of individual attitudes, beliefs, hopes, prejudices, and demands [25]. Activities, interests, and opinions (AIO) is one of the most widely used lifestyle measurement tools [23]. It encompasses variables from rational, specific, and behavioral psychology fields as well as from sociodemographic attributes, and provides good insight into the lifestyles of individuals [26]. Indeed, individuals' value priorities are part of their basic world views, and are commonly defined as desirable, trans-situational goals that vary in importance and serve as guiding principles in individuals' lives [27]. Moreover, values are broader in scope than attitudes or the types of variables contained in AIO measures, and can be an important basis for segmentation. Generally, value inventories often only contain a handful of values instead of 200 or 300 AIO items. Therefore, scholars have begun to use simple, intuitive, and elegant values as another lifestyle-related psychographic segmentation method to replace the very extensive and burdensome AIO approach [22].

Early value-related works widely used three lifestyle measurements that are simpler alternatives to AIO: value, attitude, and lifestyle (VALS); list of values (LOV); and the Rokeach Value Survey

(RVS). One of the most widely used value segmentation theories is the Stanford Research Institute's consulting values and lifestyles (VALS2) typology, which contains 35 psychographic questions and four demographic questions that link demographics and purchase patterns with psychological attitudes [28].

The RVS proposed by Rokeach (1973) [29], LOV [30] developed by the Survey Research Center at the University of Michigan, and SVS [31] developed by Schwartz et al. approximately two decades ago, are three widely used instruments to assess individual values and lifestyles. RVS requires respondents to rank 18 terminal values and 18 instrumental values. The terminal values are considered to be either self-centered or society-centered, and intrapersonal or interpersonal in focus, whereas the instrumental values are moral values and competence. LOV, proposed by Kahle, Beatty, and Homer (1986) [30], is a shorter and more easily implemented instrument which includes only nine values. SVS is a more acceptable and widely used value instrument nowadays, and suggests that there are 10 primary values organized into a circumplex [27], as shown in Figure 1. This circumplex serves as the umbrella under which the majority of individual value judgments fall. Schwartz (2012) noted the shared motivational emphases of adjacent values: power (Pow), achievement (Achiev), hedonism (Hed), stimulation (Stim), self-direction (S-D), universalism (Univ), benevolence (Benev), tradition (Trad), conformity (Conf), and security (Sec) [27]. Although the theory underlying SVS discriminates 10 values, it postulates that, at a more basic level, values form a continuum of related motivations. The Schwartz value implies that the entire set of 10 values relates to any other variable (e.g., behavior, attitude, age, etc.) in an integrated manner.

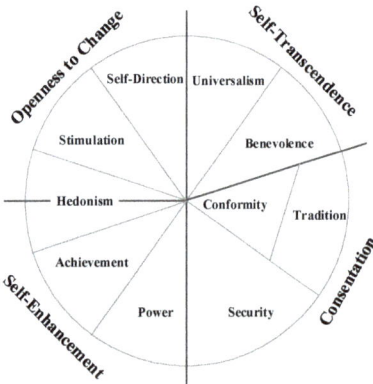

Figure 1. Schwartz Value Survey circumplex.

2.2. Psychographic Segmentation and Consumer Behavior

Market segmentation consists of viewing a heterogeneous market as a number of smaller homogeneous markets, with differing product preferences among important market segments for the purpose of developing better firm strategies. Consumer behavior and predictive preference models can be much more effective when the target audience is split into different customer segments and individually tailored predictive models are developed for each segment [32], with some of them based on psychographics.

Researchers have based the psychographic segmentation of markets on various factors, including personality, lifestyle, and values. Early psychographics mainly considered personality, but research based on one of the personality profiles, that is, Edwards Personal Preference Schedule, failed in predicting consumer purchase decisions. Other scholars used the BFF personality model. Hirsh, Kang, and Bodenhausen (2012) surveyed people about various marketing messages and found that people respond more positively to messages tailored to their personality [33]. In the music domain, I Fernández-Tobías et al. (2016) showed that online music recommendations are more successful when they leverage the correlations between people's personality and their music preferences [34]. Similarly, Karumur, Nguyen, and Konstan (2016) discovered correlations between personality and movie preferences among Netflix users [35]. Additionally, studies in marketing have found that personality explains only a surprisingly small amount of the overall variance in consumer behavior [11].

Various studies have been conducted on the effect of user lifestyles on consumers' products purchases [36] and preferences [37,38]. The comprehensive results have confirmed the effect of consumer lifestyle in predicting users' online preferences. Lee et al. (2009) assessed consumer lifestyles regarding the adoption of electronic products, and the results based on structural equation modeling demonstrated that consumer lifestyle factors (fashion consciousness, leisure orientation, internet involvement, and e-shopping preferences) are direct and indirect antecedents of consumers' intentions to adopt high-tech products [36]. Based on the AIO survey, Piazza et al. (2017) investigated the relationship between lifestyle and online user preferences, and found that there are weak monotonic correlations between Facebook likes and the lifestyles of individuals [38]. Based on a web-usage lifestyle scale developed by AIO, Pan et al. (2014) applied a tagging method to identify online lifestyles and to make recommendations according to the similarity of a user lifestyle tag to a telecommunications company product lifestyle tag [37].

In addition to lifestyle and personality, the relationship between value and consumer preferences has also been asserted. Scholars have used the LOV construct in relation to consumer fashion leadership [39], involvement [40], and brand preferences [41]. Based on questionnaire data, Weng and Run (2013) investigated the effects of Malaysian consumers' personal values using mainly LOV measurements with regard to their overall behavioral intention and purchase satisfaction [42]. Fraj and Martinez (2006) adopted VALS and identified which values and lifestyles best explain environmentally friendly behaviors [43]. Based on RVS, Padgett and Mulvey (2007) demonstrated that technology must be leveraged via links to customer values [44]. Wiedmann, Hennigs, and Siebels (2009) constructed a luxury-related value for identifying different types of luxury consumers, and found that hedonic value aspects as components of the SVS of self-directed pleasure and life enrichment were most important for their perception of luxury value [45].

The combination of segmentation and prediction, where segmentation is used to help build segment-wise prediction models, has been a common segmentation approach and has been applied to different business scenarios. Antipov and Pokryshevskaya (2010) conducted user segmentation research based on the decision tree method and then built a separate logistic regression scoring model for each segment using the churn dataset [46]. Reutterer et al. (2006) proposed a two-stage approach for deriving behaviorally persistent segments and predicting a customer's tendency to symptomatically (re)purchase using retail purchase history data [47]. Ge et al. (2011) exploited multifocal learning to divide consumers into different focal groups and automatically predict a problem category based on real-world customer problem log data [48].

2.3. Behavioral Measures of Psychographic Segmentation

Although it plays a promising role in predicting online user preferences, psychographic segmentation has not been widely studied in online shopping scenarios. There are two main reasons. First, data collection for such research typically requires online customers to complete lengthy surveys and, therefore, cannot be easily applied on a large scale. Second, acquiring the abstract psychographic characteristics of online customers through the mandatory selection of questionnaires is plagued by issues of reliability and validity for the questionnaire [16]. At the present time, with the in-depth development of big data techniques, researchers can access a large number of open-ended reports of user psychological characteristics embedded in user-generated content (UGC). In recent years, an increasing number of studies have demonstrated that such reports are ecologically valid and driven entirely by what people say they are doing and thinking, in their own words [16].

Psychology and marketing research indicate that human language reflects psychological characteristics. The frequency with which we use certain categories of words provides clues to these characteristics. Several researchers have found that variations in word use embedded in writing such as blogs, essays, and tweets can predict aspects of personality, thinking style, social connections, and purchase decisions [49–51]. These works, that explore the feasibility of deriving psychological characteristics from UGC text, have demonstrated that computational models based on derived psychological characteristics have competitive performance compared with models using a self-reported questionnaire [52,53]. Research indicates cases in which natural language data have provided a clearer picture of people's cognitive and behavioral processes than data collected from a traditional and widely used self-report survey [16].

As one of the psychographic segmentations, human values are thought to become generalized across broad swaths of time and culture, and are deeply embedded in the language that people use every day [12]. Therefore, values could be extracted from a user's online words using behavior. Boyd et al. (2015) investigated the link between user word use behavioral measurements and Schwartz value scores, and proposed the theme words associated with each SVS value dimension [16]. Regarding personality, Yarkoni (2010) reported individual words used in online blogs (N = 694) that positively and negatively correlated with personality [17]. These works constitute a proof-of-concept study that demonstrates the utility of relying on natural language markers of abstract psychological phenomena, including values and personality, and present significant opportunities to better predict and understand their connection to consumers' behaviors and thoughts in a broader sense.

2.4. Research Questions and Design

2.4.1. Q1: What Is the Predictive and Explanatory Power of Psychographic Segmentation in E-Commerce User Preferences?

Marketers and scholars have highlighted the need to account for customer perceptions and expectations in a specific market segment. In the offline context, relationships between psychographic segmentation and user behaviors and preferences have been asserted in various scenarios, and results have demonstrated that psychographic variables are effective in understanding and predicting user behaviors and preferences [26,54]. However, in the e-commerce context, the potential predictive and explanatory power between segmentations—particularly psychographic variables and the clustering method—in predicting e-commerce consumer preferences has not been comprehensively studied.

2.4.2. Q2: What Are the Differences between SVS and BFF Segmentation in Predicting and Explaining E-Commerce User Preferences?

Researchers argue that values are a useful basis for segmentation or psychographics because, compared with AIO, values are less numerous, more central and, compared with personality, more immediately related to motivations and consumer behaviors [21]. The comparison between personality and values mostly shares the context of brand preference. For example, Mulyanegara (2009; 2011) compared the predictive power of personality and values on brand preferences within a fashion context and found that values are indeed better predictors of brand preferences [55,56]. Although scholars have studied the predictive and explanatory power of psychological segmentation on consumer behavior, few studies have examined the relative importance of different psychographic variables, that is, personality and values, in understanding e-commerce consumer behavior.

2.4.3. Q3: What Is the Adaptability of Psychographic Measurements in Predicting and Explaining E-Commerce User Preferences?

As two of the psychographic measures, the SVS and BFF models have been proven to improve the understanding of consumer purchase behaviors and preferences. However, Schwartz (2012) highlighted that not all value dimensions can effectively predict individual behaviors [27]. In an online context, it is reasonable to partition the value items into more or less fine-tuned distinct values according to the needs and objectives of the analysis.

2.4.4. Overview

Psychographic segmentation has been criticized by the well-known public opinion analyst and social scientist Daniel Yankelovich, who said that psychographics is "very weak" at predicting people's purchases, making it a "very poor" tool for corporate decision-makers [57]. Studies have demonstrated that there is a significant difference between customer behaviors among different product types [58]. Different segment groups create different purchasing patterns. However, the relationship between segmentation and user buying behavior and preferences is strongly affected by the product category [42,59]. In recent years, deep learning has made great progress and has demonstrated significantly better performance compared to traditional algorithms in online big data environments. Considering the scale of online user preferences, the use of deep neural networks (DNNs) will potentially make better psychological segmentation-based preference regression models.

In this research, we introduce empirical evidence in psycholinguistic research and the recent progress of DNN algorithms in identifying and integrating consumer psychographic segments from online shopping behaviors, and investigate its performance in an e-commerce consumer preference regression model. Moreover, accounting for the predictive and explanatory power of psychographic segmentation in e-commerce consumer preferences, we use both RMSE and R^2 to conduct evaluations across five product categories. Figure 2 shows the overview of our research.

In the next section, we present the results related to the three main study questions in this research. Additionally, we investigate the promising influence of the product category and preference regression method that supports our research.

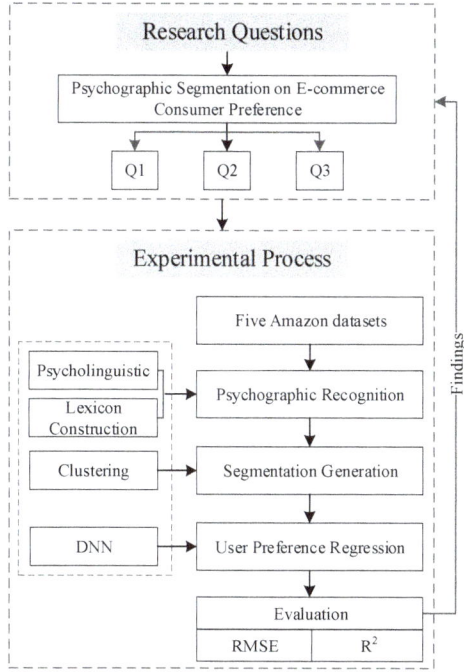

Figure 2. Research framework.

3. Methodologies

We now describe the proposed approach, beginning with intuition and then going into detail. Given that Amazon.com is the most studied online consumer domain in the context of e-commerce, we used Amazon consumer behavior data to present our methods. In this section, we introduce our methods, and apply these methods to identify Amazon customers' SVS and BFF scores to predict their online preference.

3.1. Word Use-Based Psychographic Inference

Lexicon is the basis for utilizing customer word use behaviors to automatically extract psychological traits. The combination of semantic knowledge and online corpora not only utilizes the semantic relationships as prior knowledge to provide accurate seed words but, also, help to foster the word association information, such as term position and word co-concurrency in the corpus [60]. Based on semantic knowledge and Amazon corpus, our research applies the NLP method to automatically construct SVS and BFF consumer psychographic lexicons for consumer value and personality identification. First, we obtained two seed words according to the SVS- and BFF-related word use behavior proposed by prior psycholinguistic works. Second, we applied the semantic knowledge embedded in WordNet, which is an English lexical database, to extend the seed words. Third, we further extended these seed words on the large-scale Amazon consumer review corpus, to construct the psychographic lexicons. Finally, we calculated the multidimensional psychographic scores for consumers and products based on the lexicons and consumer online reviews.

3.1.1. Construction of Psychographic Seed Words by Psycholinguistics

The empirical evidence between psychographics and consumer word use behavior is the solid basis for seed word set construction. Boyd et al. (2015) provided three value-related word use behavior sources: self-report "Value Essay", Self-Report Behavior Essay, and Facebook updates. They then presented the theme words as positive or negative using the Schwartz values [16]. Basically, we combine these three sources to build two SVS lexicon seed word sets. However, there are some contradictions, such as that Religion-related words in the Value Essay are negatively related with Self-Direction ($R^2 \geq 0.01$), but in the Self-Report Behavior Essay, these words are positively related with user Self-Direction ($R^2 \geq 0.01$)]. Considering that behavioral measures tend to outperform self-report as addressed by Boyd et al. 2015, we give priority to the user behavior measurement or the use of words in Facebook updates. Similarly, the work of Yarkoni (2010) provides the basis to construct BFF seed word sets [17]. Tables 1 and 2 summarize themes related to SVS and seed words related to BFF; see Boyd et al. 2015 for more details.

Table 1. Schwartz Value Survey (SVS)-related themes.

SVS Direction	SVS Dimensions	Positively Related Themes	Negatively Related Themes
Self-Enhancement	Achievement	Swearing	Empathy, Lord,
Self-Transcendence	Benevolence	Faith (Positive), Faith (Negative), Empathy, Honesty, Family Care, School, Meetings	
Conservation	Conformity	Faith (Positive), Faith (Negative), Family Growth, Family Care, Religiosity, Celebration, Group Success, Proselytizing	Faith (Negative), Faith (Positive)
Self-Enhancement	Hedonism	Relaxation, Swearing	Family Care, Meetings, Achievement, Religiosity, Religiosity
Self-Enhancement	Power	Student	Faith (Negative)

Table 1. Cont.

SVS Direction	SVS Dimensions	Positively Related Themes	Negatively Related Themes
Conservation	Security	Family Growth, Proselytizing	Strive, Action, Nature, Personal, Free, Band, Guitar, Rehearsal, Perform, Money, Enjoy, Spend, Free, Change
Openness to change	Self-Direction	Faith (Negative), Faith (Positive), Social, Growth, Indulgence, Caring/Knowledge	Faith (Negative), Faith (Positive), Social, Knowledge Gain, Family Care, Social, Growth, Indulgence, Caring/Knowledge
Openness to change	Stimulation	Indulgence, Knowledge Gain, Exhaustion	Faith (Positive), Faith (Negative), Family Care, Religiosity, Celebration
Conservation	Tradition	Religiosity, Celebration, Proselytizing, Faith (Positive), Faith (Negative), Family Growth, Social, Family Care	Enjoyment
Self-Transcendence	Universalism	Empathy, Faith (Positive), Faith (Negative), Empathy, Family Growth, Social, Growth, Caring/Knowledge, Openness	Daily Routine, Family Care, Student, Faith (Positive), Faith (Negative), Social, Knowledge Gain

Table 2. Big Five Factor (BFF)-related seed words.

Big Five Factors	Positively Related Words	Negatively Related Words
Neuroticism	Awful, though, lazy, worse, depressing, irony, terrible, stressful, horrible, sort, annoying, ashamed, ban	Road, Southern, visited, ground, oldest, invited, completed
Extraversion	Bar, drinks, restaurant, dancing, restaurants, grandfather, Miami, countless, drinking, shots, girls, glorious, pool, crowd, sang, grilled	Other, cats, computer, minor
Openness	Folk, humans, of, poet, art, by, universe, poetry, narrative, culture, century, sexual, films, novel, decades, ink, passage, literature, blues	giveaway
Agreeableness	Wonderful, together, visiting, morning, spring, walked, beautiful, staying, felt, share, gray, joy, afternoon, day, moments, hug, glad	Porn, cock, fuck
Conscientiousness	Completed, adventure, adventures, enjoying, Hawaii, deck	Stupid, boring, desperate, saying, utter, it's, extreme

WordNet is a large English lexical database that groups nouns, verbs, adjectives, and adverbs into sets of cognitive synonyms, each expressing a distinct concept. Synonym sets are interlinked by means of conceptual semantic and lexical relations. WordNet's structure makes it a useful tool for NLP. Based on WordNet and the psychographic seed words, we extracted the related words by following the Synonym Rule: for each positive seed word, its synonym is considered to have the same SVS or BFF dimension. Therefore, we obtain the consumer psychographic (SVS and BFF) candidate thesaurus composed of seed words and their synonyms from WordNet.

3.1.2. Psychographic Candidate Thesaurus Extension by Amazon Corpus

Consumer word use behaviors in UGC are very diverse. For one thing, the WordNet-based word extension ignores contexts where words often have different meanings [49,61]. The word "depressed", for example, can variously refer to sadness, an economic condition, or even the physical state of an object. Thus, in specific online shopping context, it is difficult to identify a potential word use characteristic using the candidate thesaurus alone.

In recent years, many scholars have adopted an NLP method called word embedding to discover new context-wise words in which a specific meaning is embedded. Based on the large-scale internet corpus, these methods have achieved remarkable progress [62]. Word embeddings map words to high-dimensional numerical vectors that contain the semantic information in these words. The basic idea of our method is to map each word into a K-dimensional vector space through word embeddings and to calculate the semantic similarity between words in corpus and words in a thesaurus using the cosine distance between vectors. Based on the psychographic candidate thesaurus, we use Word2Vec to construct word vector models which can identify new words in large-scale internet corpora by word similarity.

We apply cosine distance to measure word semantic similarity between word vectors in the psychographic candidate thesaurus and the corpus. Let $word_1 = (v_1^{w_1}, v_2^{w_1}, \ldots v_m^{w_1})$, $word_2 = (v_1^{w_2}, v_2^{w_2}, \ldots v_m^{w_2})$, where m is the word vector dimension; we use $m = 200$ in our experiments. We can then write cosine distance as follows:

$$sim(word_1, word_2) = \cos\theta = \sum_{k=1}^{m} \frac{v_k^{w_1} \times v_k^{w_2}}{\sqrt{\sum_{k=1}^{m}(v_k^{w_1})^2 \times \sum_{k=1}^{m}(v_k^{w_2})^2}}, \tag{1}$$

where the numerator represents the dot product of the two vectors and the denominator represents the modular product of the word vectors.

Based on the cosine distance, we utilize the Top10 algorithm in the Gensim [63] software implementation of Word2vec to calculate the 10 most similar words to the seed words in the thesaurus. We set 0.45 as the threshold for the similarity value according to our empirical results, and traverse the word embedding trained on Amazon corpus [64] to obtain the top 10 most similar words for the psychographic (SVS and BFF) candidate thesaurus. We then extend the candidate thesaurus by adding the top 10 most similar words. We obtain the final candidate thesaurus by carrying out the above procedure repeatedly, until there are no new words to be extracted.

Considering that consumer online word use is sometime ambiguous and that the adjacent values of psychographic are continuous, we calculate multidimensional values corresponding to the psychographics for every candidate word. According to the online psychographic score equation below, we calculate the semantic distances between each pair of words in the extended psychographic seed word sets to obtain the score of the word in all SVS or BFF dimensions. The online psychographic scores equation is

$$word_{psy\_scores}(w_{ext}, w_{psydim}) = Max\{sim(w_{ext}, w_{sed1}), sim(w_{ext}, w_{sed2}), \ldots, sim(w_{ext}, w_{sedp})\}, \tag{2}$$

where w_{ext} is a word from the final online psychographic candidate thesaurus, w_{psy_dim} is a specific psychographic dimension, w_{sed1} and $w_{sed2}, \ldots, w_{sedp}$ are seed words attached to w_{psy_dim} in the candidate thesaurus constructed in Section 3.1.1. In addition, three experts are hired to manually verify the SVS and BFF lexicon and to remove words marked as irrelevant by at least two experts. We then get positive or negative SVS and BFF lexicons with similar meanings to the seed words included.

3.1.3. Lexicon-Based Psychographic Inference

We define the p-dimensional positive/negative psychographic scores attached to an consumer/product as $L = \{L_1, L_2, \ldots, L_p\}$, user/product reviews set as $\{r_1, r_2, \ldots, r_m\}$, and the total number of a user's reviews as m and one of the review r_i as $\{w_{i1}, w_{i2}, \ldots, w_{in}\}$, where n is the total number of words in the review. According to the psychographic lexicon, we calculate consumer psychographic scores using Equation (2), where p-dimensional psychographic scores are defined as $L^{u\prime} \in L_p^{u\prime}$, where w_{ij} is the psychographic score attached to a word in the consumer reviews.

$$L^{u\prime} \in L_p^{u\prime} = \sum_{i=1}^{m}\sum_{j=1}^{n} w_{ij} \qquad (3)$$

The values are normalized as

$$L_{10}^{u} = \frac{L^{u\prime}}{Max(L_{10}^{u}{}')}. \qquad (4)$$

Overall, we calculate the positive and negative psychographic (BFF and SVS) scores for each consumer and product in the Amazon review dataset.

3.2. User Preference Prediction Based on Psychographic Segmentation and Neural Network

Based on the psychographic inference method, we obtained the psychographic scores for each consumer and product in the Amazon dataset. In this section, we further introduce density-based spatial clustering of applications with noise (DBSCAN) and a few regression methods, including DNN, to predict user preference.

3.2.1. Psychographic Segmentation Based on DBSCAN

Clustering-based segmentation is often used to categorize consumers on the basis of the relative importance they place on various product attributes, benefits, personality, and value [45,65]. DBSCAN cluster analysis does not require one to specify the number of clusters in the data a priori, as opposed to k-means. DBSCAN can find arbitrarily shaped clusters. It can even find a cluster completely surrounded by, but not connected to, a different cluster. We conduct DBSCAN on the consumers' psychographic scores, namely, the SVS and BFF scores (see Section 3.1). In the subsequent e-commerce consumer preference-predicting process, we predict the preferences based on the positive and negative consumer/product psychographic scores and the segmented groups they belong to.

3.2.2. User Preference Prediction Based on Psychographic Segmentation and Deep Neural Network

Let U and I be the number of consumers and products, R the training interaction matrix, and $\hat{R}$ the predicted interaction matrix. Let Y be a set of $y_{u,i}$, $\hat{Y}$ be a set of $\hat{y}_{ui}$, y_{ui} be the preference of user u to product i, and $\hat{y}_{ui}$ denote the corresponding predicted score. We use a partially observed vector (rows of the features set X and $\hat{X}$) as a combination of consumer representation $x^{(u)}$ and product representation $x^{(i)}$, where $x^{(u)} = \{u_{P_1}^{psy}, u_{P_2}^{psy}, \ldots, u_{P_D}^{psy}, u_{N_1}^{psy}, u_{N_2}^{psy}, \ldots, u_{N_{D\prime}}^{psy}\}$, $x^{(i)} = \{i_{P_1}^{psy}, i_{P_2}^{psy}, \ldots, i_{P_D}^{psy}, i_{N_1}^{psy}, i_{N_2}^{psy}, \ldots, i_{N_{D\prime}}^{psy}\}$. $u_{P_D}^{psy}$ and $u_{N_D}^{psy}$ represent positive and negative psychographic scores, respectively, "psy" represents psychographic methods (SVS or BFF), and P_D and $N_{D\prime}$ represent the number of dimensions of psychographic lexicons (positive or negative), respectively. We define R as $\{X, Y\}$ and $\hat{R}$ as $\{\hat{X}, \hat{Y}\}$. Then, we conduct three main segment-wise regression algorithms to build the preference-predicting models.

Random Forest (RF): RF has been shown to be effective in a wide range of segment-wise user preference predicting problems [66]. Random forest is an ensemble of binary trees $\{T_1(X), \cdots, T_B(X)\}$, where $X = x^{(u)} \cap x^{(i)} = \{x_1, \cdots, x_P\}$ is a p-dimensional vector of molecular descriptors or properties associated with a molecule. Each of these trees is stochastically trained on random subsets of the data. The ensemble produces B outputs $\{\hat{Y}_1 = T_1(X), \cdots, \hat{Y}_B = T_B(X)\}$, where $\hat{Y}_b$, $b = 1, \cdots, B$ is the prediction for a molecule by the bth tree. Outputs of all trees are aggregated to produce one final prediction. $\hat{Y}$ in regression is the average of the individual tree predictions.

Support Vector Machine (SVM): The ability of the support vector machine has been recognized in automated user/product modeling to predict retailer outcomes [67,68]. An SVM discriminates between data by constructing a hyperplane $w^T \varphi(X) + b = 0$ by minimizing $\frac{\|w\|^2}{2} + C \sum \epsilon_i$, subject to $y_i(w^T \varphi(x_i) + b) \geq 1 - \epsilon_i$, $\epsilon_i \geq 0$ $\forall i$, where $\varphi(x_i)$ is either x_i or a higher dimensional representation of x_i, C is the cost parameter, w is the weight vector, ϵ_i is margin of tolerance where no penalty is given to errors [69].

Deep Neural Network (DNN): Thanks to deep learning's (DL) capability in solving many complex tasks while providing state-of-the-art results, academia and industry alike have been in a race to apply deep learning to a wider range of applications [70]. Although neural networks are widely used in market segmentation [71,72], few researchers have introduced deep neural networks (DNNs) into market segmentation to predict user preferences. DNN enables effective capture of non-linear and non-trivial user-product relationships. A typical DNN consists of an input layer, a few hidden layers, and an output layer with number of nodes equal to the cardinality of categories. We use the typical mean square error (MSE), i.e., $\sum_{i=1}^{B}(y_i - \hat{y}_i)^2$ as a loss function of our DNN method, where $\hat{y}_i = h(g(x_i))$, g is some linear combination of node values in the hidden layers of the network, and h is an activation function (typically sigmoid or hyperbolic tangent function). Training a neural network involves minimizing the loss function defined above using gradient descent and backpropagation of the gradient across multiple hidden layers to update the weights between nodes. Here, we initialize the weight using Gaussian distribution with expectation 0 and variance 1, then form a neural network with two hidden layers for consumer preference regression tasks.

3.3. Evaluation

RMSE and R^2 are two of the most widely used consumer preference evaluation methods for rating prediction and can reflect the performance of psychographic segmentations in user preference and rating prediction. We apply both in an evaluation method for the prediction. Let $\hat{y}_{u,i}$ be the predicted ratings, $y_{u,i}$ the true ratings, $\overline{y_u}$ the mean rating for consumer u, and n the size of test dataset. Then, RMSE and R^2 are defined as follows:

$$\text{RMSE} = \sqrt{\frac{1}{n} \sum_{i=1}^{n} (y_{u,i} - \hat{y}_{u,i})^2}, \tag{5}$$

$$R^2 = \frac{\sum_{i=1}^{n}(\hat{y}_{u,i} - \overline{y_u})^2 - \sum_{i=1}^{n}(y_{u,i} - \hat{y}_{u,i})^2}{\sum_{i=1}^{n}(y_{u,i} - \overline{y_u})^2}. \tag{6}$$

4. Experiment

In the previous section, we have constructed the positive and negative e-commerce psychographic lexicons, namely SVS-pos, SVS-neg and BFF-pos, BFF-neg lexicons (see S1.csv, S2.csv, S3.csv and S4.csv respectively in supplementary materials for more details), and conducted e-commerce consumer segmentation based on the identified SVS and BFF scores and DBSCAN. In this section, we further proposed a DNN method to build the segment-wise consumer rating regression model. Then, we proceeded to utilize the online shopping data in Amazon to conduct our experiments.

4.1. Dataset Description

Amazon is one of the largest e-commerce platforms in the world and has accumulated a large amount of user buying behavior data. The Amazon review dataset, published by McAuley et al. [64], contains product reviews and metadata from Amazon.com, including 142.8 million reviews spanning from May 1996 to July 2014. We selected 5 review datasets from 5 product categories according to "K-core" values of "10", whereby each of the remaining users or items have at least 10 reviews.

According to the work by Arnoux et al. (2017) [73], we consider that 10 reviews (whereby the average length of review is 189 words) is capable for consumer/product psychographic inference and comparable to 25 tweets. Table 3 shows the detailed dataset description.

Table 3. Experimental dataset description.

Item Category	Total Number of Users	Total Number of Items	K-Core
Beauty	1397	548	10
Office Products	724	395	10
Baby	946	231	10
Grocery and Gourmet Food	1556	494	10
Toys and Games	1037	398	10

The sample review shows details about our data:
{
"reviewerID ': 'A2SUAM1J3GNN3B",
"asin': '0000013714",
"reviewerName': 'J. McDonald",
"helpful': [2,3],
"reviewText': 'I bought this for my husband who plays the piano. He is having a wonderful time playing these old hymns. The music is at times hard to read because we think the book was published for singing from more than playing from. Great purchase though!",
"overall": 5.0,
"summary': 'Heavenly Highway Hymns",
"unixReviewTime": 1252800000,
"reviewTime': '09 13, 2009"
}

4.2. Experimental Procedure

The data processing process is divided into the following steps. Figure 3 shows the whole picture of our experiment process across Sections 3 and 4.

First of all, we keep the "reviewerID", "asin", "overall", "reviewText", and "summary" tags in the seven datasets above and combined "reviewText" and "summary" as the word use behaviors in recognized online psychographics.

Second, we conduct text preprocessing, including normalization, tokenization, removing stop words, and stemming on the textual contents using Amazon reviews and lexicons by Python, machine learning tool Scikit-learn (https://scikit-learn.org/stable/modules/generated/sklearn.cluster.DBSCAN.html) and the Natural Language Toolkit (NLTK) [74]. Normalization is a process that converts a list of words to a more uniform sequence. Given a character sequence and a defined document unit, tokenization is the task of chopping it up into pieces, called tokens. Some extremely common words in reviews, which would be of little value in helping select texts matching our need, are excluded from the vocabulary entirely. These words are called stop words. The goal of stemming is to reduce inflectional forms, and sometimes derivationally related forms of a word, to a common base form. We perform stemming by Lancaster Stemmer in NLTK on both words in lexicon and reviews. We removed stop words using the English stop word list in NLTK, obtained the of words in lowercase using the text lower method in Python, and conducted Z-score normalization using scale method in Scikit-learn. Based on the SVS-pos, SVS-neg, BFF-pos, and BFF-neg lexicons in Section 3.1, and all the data preprocessing steps above, we calculate the psychographic scores by matching the used words and the words in these lexicons. Thus, we get the SVS and BFF scores for each Amazon consumer and product.

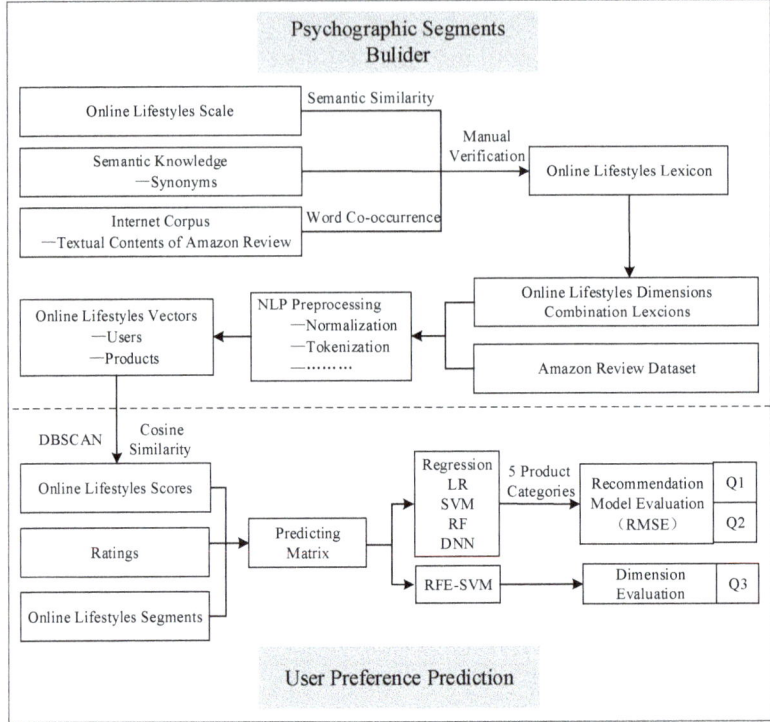

Figure 3. Experiment process.

Third, for each of the product categories, we perform the DBSCAN algorithm in the consumers' SVS or BFF scores, using Scikit-learn, to get the consumers' positive and negative psychographic scores and attached psychographic segment tags. We then build the rating predicting dataset which combines psychographic scores (as independent variables) with the rating given by the consumer to the product (as a dependent variable). We also construct feature sets for each product category which contains random values between 0 and 1 as the control group. For each of these datasets, we optimize the parameters for DNN by gradient descent. The optimal number of epochs (a single pass through the full training set) used for the neural network algorithm is decided by the performance on the validation set. For SVM, we use the validation set for optimizing the cost parameter C. We use Scikit-learn (https://scikit-learn.org/stable/modules/generated/sklearn.linear_model.LinearRegression.html) implementation of linear regression (LR), SVM with a radial basis function kernel, random forest, and Keras implementation of Google TensorFlow software (https://www.tensorflow.org) for developing DNN and baseline. A 5-fold cross-validation is applied to select training and test datasets for each fold and avoid overfitting for linear regression, SVM, RF, and DNN. An example of RMSE evolution with epochs for DNN is shown in Figure 4. In Figure 4, we can see that the best epoch is around 15.

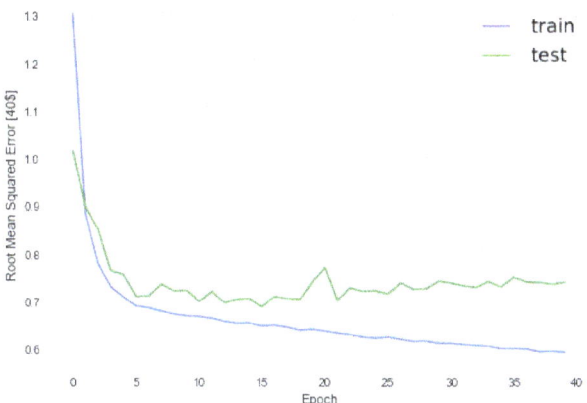

Figure 4. Deep neural network training and testing with different epochs.

Fourth, we apply feature ranking with a recursive feature elimination method to investigate the feature importance of different psychographic subdimensions in understanding consumer online preference. The support vector machine-based recursive feature elimination (RFE-SVM) approach is a popular technique for feature selection and subsequent regression task, especially in predicting consumer preference [75]. At each iteration, a linear SVM is trained, followed by removing one or more "bad" features from further consideration. The quality of the features is determined by the absolute value of the corresponding weights used in the SVM. The features remaining after a number of iterations are deemed to be the most useful for discrimination and can be used to provide insights into the given data [76]. By introducing the RFE-SVM in segment-wise consumer rating prediction, we can dive into the subdimensions of SVS and BFF to explore whether these dimensions are effective in predicting and explaining preferences.

Finally, we perform an online consumer preference-predicting experiment that contains 5 product categories * 4 predictive algorithms (LR, SVM, RM, NN) * 3 psychographic variables tools (Random, SVS, and BFF) * 2 clustering method (clustering consumers based on DBSCAN or not).

5. Result Analysis

5.1. Analysis of the Significance of Different Experimental Settings in Predicting and Explaining E-Commerce User Preferences

We evaluate the predictive and explanatory power of user preference prediction based on LR, support vector machine (SVM), random forest (RF), and DNN, integrating consumer psychographic segmentation across different Amazon product categories. Table 4 shows part of the results of RMSE and R^2 under different experimental settings for the user preference prediction process. The nine columns are "ProductCatrgory", "ClusterId", "SampleNum", "Psychographic Variable", "Clustering or Not", "RegressionMethods", "RMSE", "R^2" and "ReviewLen". The "ProductCatrgory" column contains five Amazon product categories from Table 3; "ClusterId" is the tag for each segment; "SampleNum" is the number of Amazon consumer ratings and reviews; "Psychographic Variables "; " Psychographic Variables" is the psychographic variable that we use to perform consumer segmentation; "Clustering or Not" is the experiment setting of applied DBSCAN clustering that determines whether consumer segmentation is performed; "RegressionMethods" is the regression model used in preference prediction; and "RMSE" and "R^2" are the RMSE and R^2, respectively, for regression model evaluation; "ReviewLen" is the average amount of words in review for each dataset. We perform analysis of variance (ANOVA) on the results to test whether different experimental settings (product category, predictive algorithms, psychographic variables, and clustering or not) have an influence on the performance (R^2 and RMSE) of preference prediction.

Table 4. Evaluation results of user preference prediction under different experimental settings (partial part of).

Product Category	ClusterId	SampleNum	Psychographic Variables	Clustering or Not	RegressionMethods	RMSE	R^2	ReviewLen
Toys and Games	Random	9577	Random	Random	RF	0.9544	−0.0462	105.3555
Toys and Games	Random	9577	Random	Random	RF	0.9532	−0.0434	105.3555
Toys and Games	Random	9577	Random	Random	LR	0.9362	−0.0065	105.3555
Toys and Games	Random	9577	Random	Random	LR	0.9351	−0.0042	105.3555
Toys and Games	All	9577	BFF	Not Clustered by DBSCAN	SVR	0.9091	0.0509	105.3555
Toys and Games	All	9577	BFF	Not Clustered by DBSCAN	LR	0.9084	0.0524	105.3555
Toys and Games	All	9577	SVS	Not Clustered by DBSCAN	SVR	0.9067	0.0558	105.3555
Toys and Games	All	9577	SVS	Not Clustered by DBSCAN	LR	0.9007	0.0684	105.3555
Toys and Games	Cluster_2	2044	SVS	Clustered by DBSCAN	NN	1.1217	0.0311	105.3555
Toys and Games	Cluster_2	2044	SVS	Clustered by DBSCAN	RF	0.9350	0.0746	105.3555
Digital Music	Cluster_1	3920	SVS	Clustered by DBSCAN	SVR	1.1081	−0.0632	205.5029
Digital Music	Cluster_1	3920	SVS	Clustered by DBSCAN	RF	1.0575	0.0317	205.5029

Table 5 shows the ANOVA results which demonstrate that the product category, clustering method, and regression methods have a significant effect on both R^2 and RMSE (p-value < 0.000). For the segmentation variables that we use (SVS, BFF, and random segmentation), there are significant differences between the three groups (p-value < 0.05) for R^2, whereas the difference is not significant for RMSE.

Table 5. ANOVA results for experimental settings in predicting consumer rating.

Index	Evaluation	F	PR (>F)	df	Sum_sq
Psychographic	R^2	36.39	0.0000	2	0.4120
Regression Methods	R^2	32.55	0.0000	3	0.5526
Product Category	R^2	7.44	0.0000	4	0.1684
Residual	R^2			147	0.8320
Clustering or Not	R^2	35.20	0.0000	2	0.4028
Psychographic	RMSE	9.93	0.0001	2	0.0516
Clustering or Not	RMSE	13.17	0.0000	2	0.0659
Regression Methods	RMSE	102.35	0.0000	3	0.7677
Product Category	RMSE	76.31	0.0000	4	0.7632
Residual	RMSE			147	0.3676

5.2. Analysis of the Effect of Psychographic Segmentation on Predicting and Explaining E-Commerce User Preferences

To investigate our three main questions, we further compare the differences between the SVS, BFF, and random segmentation, and their subdimensions, in predicting and explaining consumer preferences, while excluding the influence of other experimental settings.

5.2.1. Q1: Analysis of the Effect of the Clustering-Based Segmentation in Predicting and Explaining User Preferences

We conduct Tukey's range test, which is a single-step multiple comparison procedure and statistical test, on Table 4. Table 6 shows the Tukey's range test results for the preferences based on the psychographic variables and DBSCAN segmentation method. Surprisingly, we can see that there is no significant difference in both RMSE and R^2, regardless of whether we use DBSCAN to obtain psychographic segments. Additionally, the psychographic variables significantly improve the performance of user preference explanation (R^2), whereas the improvement of efficiency is not significant in user preference prediction (RMSE).

Table 6. Tukey's range test for psychographic segmentation as a whole in understanding consumer preference.

	Tukey's Test for Segmentation						
Evaluations	G 1	G 2	Lower	Meandiff (G2−G1)	Upper	Reject (p < 0.05)
R^2	Clustered by DBSCAN	Not Clustered by DBSCAN	−0.05	−0.0027	0.0446	False
R^2	Clustered by DBSCAN	Random	−0.167	−0.1201	−0.0732	True
R^2	Not Clustered by DBSCAN	Random	−0.1721	−0.1174	−0.0626	True
RMSE	Clustered by DBSCAN	Not Clustered by DBSCAN	−0.0623	−0.011	0.0403	False
RMSE	Clustered by DBSCAN	Random	−0.0103	0.0406	0.0914	False
RMSE	Not Clustered by DBSCAN	Random	−0.0078	0.0516	0.1109	False

5.2.2. Q2: Analysis of the Significance and Differences between Psychographic Tools in Predicting and Explaining User Preferences

We applied Tukey's range test to further compare the differences between the SVS, BFF, and random segmentation within psychographic variables. Table 7 shows the results. For all product categories that we study, there are no significant differences between the three groups in terms of RMSE. By contrast, there are significant differences between both the SVS and BFF with the random segmentations in R^2.

Table 7. Tukey's test for different psychographic variables.

Evaluations	G1	G2	Lower	Meandiff (G2-G1)	Upper	Reject ($p < 0.05$)
RMSE	BFF	Random	−0.0145	0.0393	0.0932	False
RMSE	BFF	SVS	−0.052	−0.0037	0.0445	False
RMSE	Random	SVS	−0.0964	−0.0431	0.0102	False
R^2	BFF	Random	−0.158	−0.1091	−0.0602	True
R^2	BFF	SVS	−0.0196	0.0242	0.0681	False
R^2	Random	SVS	0.0849	0.1333	0.1817	True

We plot the mean values of RMSE and R^2 for the three segmentation variables in the different product categories; Figure 5 shows the results. For differences between the SVS and BFF in user rating explanation, the SVS is superior to BFF in explaining consumer preferences (R^2) across all product categories, except for Toys and Games. For user rating prediction, the SVS and BFF tend to perform better than random segmentation in predicting user preferences (RMSE) in Beauty, Digital Music, and Office Products. Compared with BFF, the SVS is slightly more powerful in predicting user ratings (RMSE) for Beauty and Digital Music, whereas there is no difference between them for Grocery and Gourmet Food, and Office Products. However, the SVS was useless in predicting user ratings (RMSE) for Toys and Games, whereas BFF improved the predictive power of user ratings.

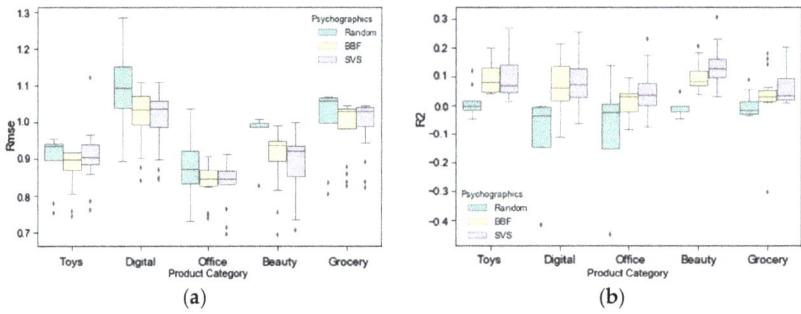

Figure 5. Mean values of RMSE and R^2 about three segmentation variables across different product Categories. (a) Mean values of RMSE; (b) Mean values of R^2.

5.2.3. Q3: Analysis of the Adaptability of Psychographic Measurement in Predicting and Explaining E-Commerce User Preferences

In Section 5.2.2, we studied the predictive and explanatory power of psychographic segmentation across different product categories. However, it remains unclear whether all subdivision dimensions of psychographic variables help understand user preferences across different e-commerce product categories. Table 8 shows part of the results of different subdimensions' psychographic variables in predicting and explaining user preferences using the RFE-SVM approach. In Table 8, feature ranking corresponds to the ranking position of the i_{th} psychographic subdimension, where the selected, i.e., estimated best dimensions are assigned rank 1; "importance support" represents these selected dimensions (with value TRUE); "feature source" represents whether a dimension is positively or negatively related to the SVS or BFF; "product category" contains five products from Table 8; "psychographic" contains three psychographic segmentations including the SVS, BFF, and random segmentation; and "subdimension" represents the subdimensions of positive and negative SVS and BFF. There are 810 rows in Table 8, and 150 rows with psychographic subdimensions that support our rating prediction models. We group the importance ranking in the 150 rows by product category and plot the total rankings for different subdimensions of the SVS and BFF. Only approximately

18.5% of the subdimensions of the psychographic measurements were effective in understanding specific e-commerce preferences. We can see that within Table 8, with the exception of Sec and Univ, all subdimensions of the SVS and BFF are useful. However, in different product categories, there are more subdimensions that are not effective in understanding consumer preferences, including Benev, Hed, Trad, Openness, and Extraversion in Beauty; Benev, Conf, and Openness in Office Products; Benev, Hed, Trad, Openness, and Extraversion in Toys and Games; Achiev, Hed, and Stim in Digital Music; and Achiev, Benev, Conf, and Openness in Grocery and Gourmet Food.

Table 8. Importance of different subdimensions of psychographic variable in predicting and explaining user preference.

Importance Ranking	Importance Support	Feature Source	Product Category	Psychographic	Subdimension	Consumer or Product
8	FALSE	pos	Digital Music	BFF	Openness	U
1	TRUE	pos	Digital Music	BFF	Agreeableness	U
1	TRUE	pos	Digital Music	BFF	Conscientiousness	U
6	FALSE	pos	Digital Music	BFF	Neuroticism	I
5	FALSE	pos	Digital Music	BFF	Extraversion	I
3	FALSE	pos	Digital Music	BFF	Openness	I
...	...	...	Digital Music	...	...	...
7	FALSE	pos	Digital Music	SVS	Stim	U
26	FALSE	pos	Digital Music	SVS	Conf	U
25	FALSE	pos	Digital Music	SVS	Hed	U
5	FALSE	pos	Digital Music	SVS	Trad	U
14	FALSE	pos	Digital Music	SVS	Achiev	U
17	FALSE	pos	Digital Music	SVS	Benev	U
4	FALSE	pos	Digital Music	SVS	Pow	U
13	FALSE	pos	Digital Music	SVS	Univ	U
16	FALSE	pos	Toys and Games	SVS	S-D	I
29	FALSE	pos	Toys and Games	BFF	Sec	I
1	TRUE	pos	Toys and Games	BFF	Stim	I
9	FALSE	pos	Toys and Games	BFF	Conf	I
24	FALSE	pos	Toys and Games	BFF	Hed	I
19	FALSE	pos	Toys and Games	SVS	Trad	I
22	FALSE	pos	Toys and Games	SVS	Achiev	I
30	FALSE	pos	Toys and Games	SVS	Benev	I

We conduct Tukey's range test based on the 150 rows with psychographic subdimensions supporting our regression models; Table 9 shows the results. There are 37 significant comparisons that indicate the relationship between subdimensions of the SVS and BFF. From Table 9, we can see that, in terms of the effectiveness of psychographic subdimensions in understanding consumer preferences, the SVS subdimensions tend to follow the descending order of Pow, Trad, Conf, SD, Benev, and Stim. For subdimensions of BFF, the descending order is Conscientiousness, Neuroticism, and Openness. Figures 6 and 7 show two intuitive expressions of these orders, respectively.

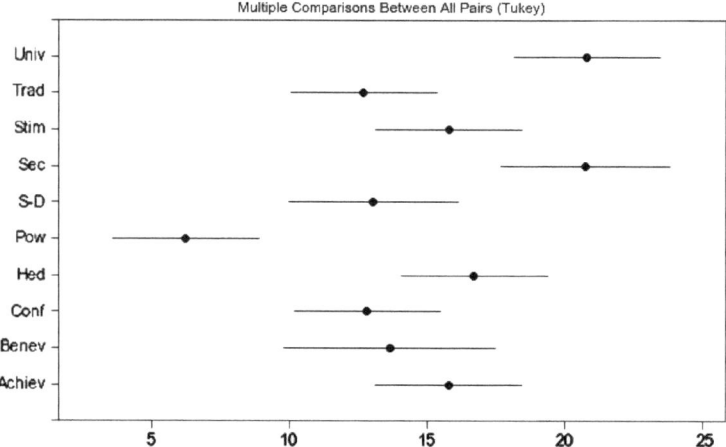

Figure 6. Difference between subdimensions of SVS in understanding consumer rating.

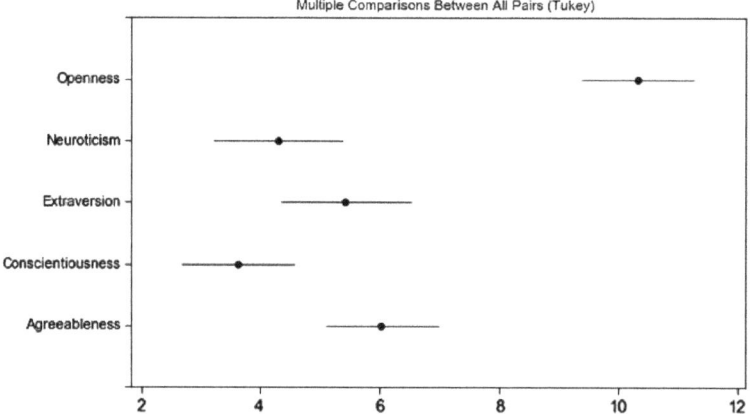

Figure 7. Difference between subdimensions of BFF in understanding consumer rating.

Table 9. Difference between subdimensions of psychographic in understanding user preference.

Ps1–Ps2 (Importance Ranking)	Diff	Lwr	Upr	P.adj
Subdimensions of SVS				
Pow–Achiev	−9.5667	−14.4328	−4.7006	0.0000
Pow–Hed	−10.4833	−15.3494	−5.6172	0.0000
Sec–Pow	14.4780	9.2220	19.7340	0.0000
Stim–Pow	9.5667	4.7006	14.4328	0.0000
Univ–Pow	14.5667	9.7006	19.4328	0.0000
Univ–Trad	8.1167	3.2506	12.9828	0.0000
Univ–Conf	7.9833	3.1172	12.8494	0.0000
Trad–Sec	−8.0280	−13.2840	−2.7720	0.0000
Sec–Conf	7.8946	2.6386	13.1506	0.0000
Univ–S-D	7.7776	2.5216	13.0336	0.0001
Sec–S-D	7.6889	2.0700	13.3078	0.0004

Table 9. Cont.

Ps1–Ps2 (Importance Ranking)	Diff	Lwr	Upr	P.adj
Pow–Conf	−6.5833	−11.4494	−1.7172	0.0005
Trad–Pow	6.4500	1.5839	11.3161	0.0007
S-D–Pow	6.7891	1.5331	12.0451	0.0012
Pow–Benev	−7.4661	−13.4258	−1.5064	0.0021
Univ–Benev	7.1006	1.1408	13.0603	0.0049
Sec–Benev	7.0118	0.7297	13.2940	0.0131
Univ–Stim	5.0000	0.1339	9.8661	0.0371
Univ–Achiev	5.0000	0.1339	9.8661	0.0371
Subdimensions of BFF				
Openness–Conscientiousness	6.7000	1.8339	11.5661	0.0003
Openness–Neuroticism	6.0498	0.7938	11.3058	0.0084

5.3. Analysis of Product Category and Regression Methods in Predicting and Explaining E-Commerce User Preferences

In this section, we analyze the internal differences in the new segment-wise preference regression methods and product categories that support the main research questions. Based on Table 4, we perform the Tukey's range test on the preference prediction method and product categories; the details are shown in Table 10.

Regarding RMSE, there were more significant differences in RMSE between different product categories compared to those in R^2, which is shown in Figure 8. The rating prediction performance of psychographics tends to follow a descending order from Office Products, which obtains the best predictive power, to Toys and Games, Beauty, Grocery and Gourmet Food, and Digital Music. Regarding R^2, as shown in Figure 9, the psychographic variables in Beauty, and Toys and Games obtain a significantly greater R^2 than that of Office Products (p-value < 0.05). In terms of the regression method, as shown in Figures 10 and 11, DNN is significantly better than the other three methods in terms of both RMSE and R^2. The performance of methods tends to follow a descending order from DNN to RF, LR, and SVR in the performance of preference prediction.

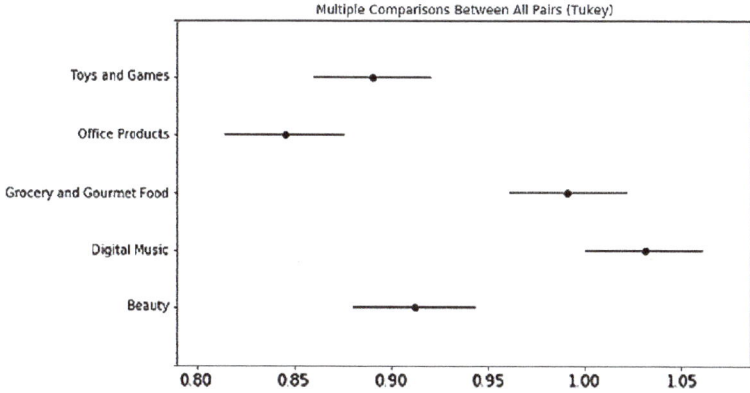

Figure 8. Multiple comparisons between all pairs (Tukey) in product categories for psychographic variable segmentation-based rating prediction (RMSE).

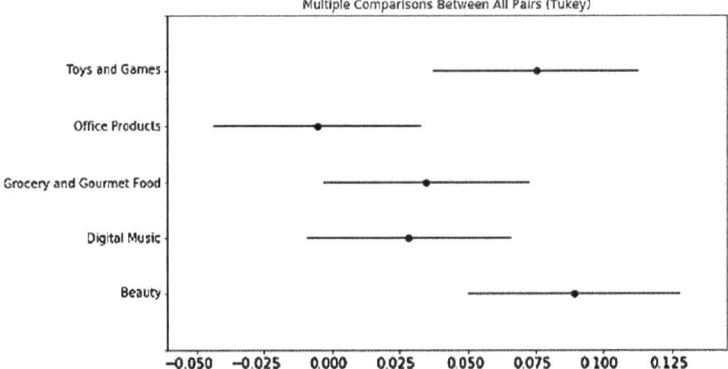

Figure 9. Multiple comparisons between all pairs (Tukey) in product categories for psychographic segmentation-based rating explaining (R^2).

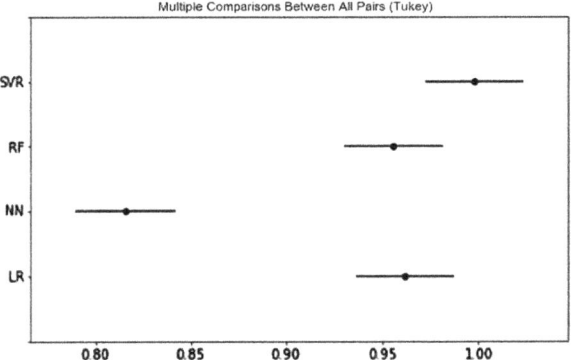

Figure 10. Multiple comparisons between all pairs (Tukey) in regression methods for psychographic segmentation-based rating prediction (RMSE).

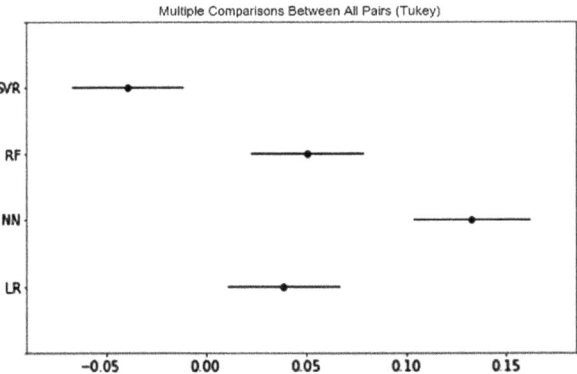

Figure 11. Multiple comparisons between all pairs (Tukey) in regression methods for psychographic segmentation-based rating explanation (R^2).

Table 10. Tukey's range test for product category and regression methods.

		Tukey's Test for Product Category				
Evaluations	g1	g2	Lower	Meandiff (g2−g1)	Upper	Reject (p Value < 0.05)
RMSE	Beauty	Digital Music	0.057	0.119	0.181	True
RMSE	Beauty	Grocery and Gourmet Food	0.017	0.079	0.142	True
RMSE	Beauty	Office Products	−0.130	−0.067	−0.004	True
RMSE	Digital Music	Office Products	−0.248	−0.186	−0.124	True
RMSE	Digital Music	Toys and Games	−0.202	−0.141	−0.079	True
RMSE	Grocery and Gourmet Food	Office Products	−0.208	−0.146	−0.084	True
R^2	Grocery and Gourmet Food	Toys and Games	−0.162	−0.101	−0.040	True
R^2	Beauty	Office Products	−0.172	−0.095	−0.017	True
R^2	Office Products	Toys and Games	0.004	0.081	0.157	True
		Tukey's Test for Regression Methods				
RMSE	LR	NN	−0.199	−0.146	−0.094	True
RMSE	NN	RF	0.088	0.140	0.192	True
RMSE	NN	SVR	0.130	0.182	0.235	True
R^2	LR	NN	0.036	0.094	0.151	True
R^2	LR	SVR	−0.135	−0.078	−0.022	True
R^2	NN	RF	−0.140	−0.082	−0.025	True
R^2	NN	SVR	−0.229	−0.172	−0.115	True
R^2	RF	SVR	−0.146	−0.090	−0.034	True

6. Discussion

In this study, we have focused on the promising role that different psychographic segmentations play in the understanding of e-commerce consumer preferences. Based on real-world user behavior data from Amazon, we have introduced psychographic-related behavioral evidence from psycholinguistics and applied NLP, clustering, and DNN methods to identify users' psychographic segments and to further predict their preferences.

We summarize our results as follows. First, we have found that dividing e-consumers into heterogeneous groups using a clustering method did not significantly improve the predictive and explanatory power of e-commerce consumer preferences. By contrast, psychographic variables significantly improved the explanatory power of e-consumer preferences, whereas the improvement in predictive power was not significant. This finding is consistent with past studies [11,42] which showed that individual segments based on their psychographic measures do not seem to provide a great deal of predictive power in the context of buying behavior.

Second, we have found that both value and personality segmentations significantly improve user preference explanation under different e-commerce scenarios, whereas no significant improvement was shown in user preference prediction. These findings have verified previous research that psychographic variables do not seem to provide substantial predictive power in the context of offline buying behavior [11]. However, these findings somehow contradict works suggesting that customer segmentation based on these variables may be easy to understand, but may not provide the best possible explanatory power [42,58]. These findings show that psychographic variables may play a more important role in understanding shopping behaviors in online, rather than offline, shopping scenarios. Additionally, although there is no significant difference between the SVS and BFF across all product categories, both the SVS and BFF tend to predict e-consumer preferences better across most of the product categories that we have studied, and the SVS seems to outperform BFF in all the product categories that we have studied, except Toys and Games. Values that characterize human motivation may be a better psychographic variable compared with personality, which emphasizes individual differences.

Third, we have found that only 18.5% of the subdimension combinations within a psychographic tool have stable effectiveness in understanding e-commerce preferences related to a specific product category. However, all BFF subdimensions are important in understanding e-consumer preferences, whereas the security and benevolence of the SVS do not demonstrate their effectiveness. We have found that using more subdimensions does not significantly improve the predictive and explanatory power within different product categories. Although the SVS is capable of explaining e-consumer preferences, our work indicates that there may be some subdimensions of the SVS that are able to perform better in understanding e-consumer preferences. This finding is consistent with Schwartz's argument that it is reasonable to partition the value items into more or less fine-tuned distinct values according to the needs and objectives of the analysis [27].

Finally, the DNN method that we have proposed obtained the best predictive and explanatory power in understanding e-consumer preferences; it is significantly better than RF and SVM that were applied in previous research. Regarding product categories, there are more significant differences for psychographic variables in predicting than explaining e-consumer preferences. Moreover, the role of psychographic variables in predicting and explaining e-consumer preferences typically does not demonstrate consistency in different product categories: psychographic variables demonstrates the best predictive power in Office Products and the least explanatory power in Office Products. In most cases, prediction models are likely to possess some level of both explanatory and predictive power [77]. We can, therefore, visualize performances of different models in terms of a trade-off between explanatory and predictive power on a two-dimensional plot, where the two axes indicate prediction accuracy and explanatory power, respectively. We leave such visualization for future work. As the influence of psychographic variables on user preferences is moderated by product category, our findings indicate a more complicated relationship between them.

This study has extended the depth and breadth of psychographic-related studies through user preference prediction in real-world e-commerce scenarios. Our findings regarding psychographic segmentation and segment-wise preference prediction have provided theoretical guidance for psychographic variable adoption, selection, and use in electronic marketing researches like online advertising, retail, and recommendation. Practically, our findings regarding subdimensions of psychographic variables have provided a practical reference for psychographic measurement development in each e-commerce product category that we have studied. Additionally, the influence of product category on psychographic-based preference prediction and explanation indicates promising e-commerce product research directions and applications. By introducing psychographic-related word use behavioral evidence, followed by natural language processing and DNN techniques, we have attempted to overcome the difficulties of observing e-consumer psychographics on a large scale, and have provided a promising psychographic-based consumer preference prediction method for subsequent research and applications.

However, our study has some limitations. First, we have only applied the review dataset with "K-core" values of 10, whereas there are a huge number of consumers who have either a limited number of reviews or words in their reviews, potentially causing bias in the psychographic inference. Second, the original dataset does not provide demographic information and we have not evaluated the difference between the psychographic scores and the scores assessed by self-report, which may have caused biases in our research results. Third, in addition to psychographic-related single words, there may be other linguistic clues embedded in phrases, sentences, and paragraphs that we have not taken into consideration. Fourth, although our research has demonstrated the significant explanatory power of psychographic tools in understanding e-consumer preferences, no significant predictive power or difference was found. Psychographic variables such as AIO, VALS, and LOV, or their combinations, should be taken into consideration in fully assessing the influence of psychographics on the understanding of e-consumer preferences.

In future studies, we will improve the consumer diversity of Amazon data to verify our findings. Promising future research directions include: evaluating the current psychographic inference method; developing a new psychographic variable identification method by introducing more advanced DNN models and textual features like SVS-related themes [16] and word embeddings [78]; verifying and comparing word use behaviors (K-core); combining different psychographic variables, together with other well-known segmentation approaches (e.g., demographic segmentation, behavioral segmentation, geographic segmentation, etc.) to further understand their influence in predicting and understanding e-consumer preferences.

Supplementary Materials: The following are available online at http://www.mdpi.com/2076-3417/9/10/1992/s1, Table S1–S4.

Author Contributions: The study was carried out in collaboration between all authors. H.L., Y.H., X.H. and W.W. designed the research topic. Y.H. and H.L. conducted the experiment and wrote the paper. Z.W. and K.L. examined the experimental data and checked the experimental results. All authors agreed to submission of the manuscript.

Funding: This research is supported by Program of National Natural Science Foundation of China (No. 71571084, No. 71271099) and China Scholarship Council.

Conflicts of Interest: The authors declare no conflict of interest.

References

1. Jacobs, B.J.D.; Donkers, B.; Fok, D. Model-based purchase predictions for large assortments. *Marketing Sci.* **2016**, *35*, 389–404. [CrossRef]
2. Lu, S.; Xiao, L.; Ding, M. A video-based automated recommender (VAR) system for garments. *Marketing Sci.* **2016**, *35*, 484–510. [CrossRef]
3. Xie, K.L.; Zhang, Z.; Zhang, Z. The business value of online consumer reviews and management response to hotel performance. *Int. J. Hospit. Manag.* **2014**, *43*, 1–12. [CrossRef]
4. Chen, J.; Haber, E.; Kang, R.; Hsies, G.; Mahmud, J. Making use of derived personality: The case of social media ad targeting. In Proceedings of the Ninth International AAAI Conference on Web and Social Media, Oxford, UK, 26–29 May 2015.

5. Trusov, M.; Ma, L.; Jamal, Z. Crumbs of the cookie: User profiling in customer-base analysis and behavioral targeting. *Marketing Sci.* **2016**, *35*, 405–426. [CrossRef]
6. Culotta, A.; Cutler, J. Mining brand perceptions from twitter social networks. *Marketing Sci.* **2016**, *35*, 343–362. [CrossRef]
7. Jin, J.; Ji, P.; Gu, R. Identifying comparative customer requirements from product online reviews for competitor analysis. *Eng. Appl. Artif. Intell.* **2016**, *49*, 61–73. [CrossRef]
8. Matz, S.C.; Netzer, O. Using big data as a window into consumers' psychology. *Curr. Opin. Behave. Sci.* **2017**, *18*, 7–12. [CrossRef]
9. Jih, W.J.K.; Lee, S.F. An exploratory analysis of relationships between cellular phone uses' shopping motivators and lifestyle indicators. *J. Comp. Inf. Syst.* **2004**, *44*, 65–73.
10. Ko, E.; Kim, E.; Taylor, C.R.; Kim, K.H.; Kang, J.I. Cross-national market segmentation in the fashion industry: A study of European, Korean, and US consumers. *Int. Marketing Rev.* **2007**, *24*, 629–651. [CrossRef]
11. Sandy, C.J.; Gosling, S.D.; Durant, J. Predicting consumer behavior and media preferences: The comparative validity of personality traits and demographic variables. *Psychol. Market.* **2013**, *30*, 937–949. [CrossRef]
12. Pennebaker, J.W.; Chung, C.K.; Frazee, J.; Lavergne, G.M.; Beaver, D.I. When small words foretell academic success: The case of college admissions essays. *PLoS ONE* **2014**, *9*, e115844. [CrossRef] [PubMed]
13. Buettner, R. Predicting user behavior in electronic markets based on personality-mining in large online social networks. *Electr. Marketing* **2017**, *27*, 247–265. [CrossRef]
14. Lee, A.J.T.; Yang, F.-C.; Chen, C.-H.; Wang, C.-S.; Sun, C.-Y. Mining perceptual maps from consumer reviews. *Decis. Support Syst.* **2016**, *82*, 12–25. [CrossRef]
15. Wang, Y.; Lu, X.; Tan, Y. Impact of product attributes on customer satisfaction: An analysis of online reviews for washing machines. *Electr. Commerce Res. Appl.* **2018**, *29*, 1–11. [CrossRef]
16. Boyd, R.L.; Wilson, S.R.; Pennebaker, J.W.; Kosinski, M.; Stillwell, J.D.; Michaela, R. Values in words: Using language to evaluate and understand personal values. In Proceedings of the ICWSM, Oxford, UK, 26–29 May 2015; pp. 31–40.
17. Yarkoni, T. Personality in 100,000 words: A large-scale analysis of personality and word use among bloggers. *J. Res. Personal.* **2010**, *44*, 363–373. [CrossRef]
18. Smith, W.R. Product differentiation and market segmentation as alternative marketing strategies. *J. Marketing* **1956**, *21*, 3–8. [CrossRef]
19. Karlsson, L.; Dolnicar, S. Someone's been sleeping in my bed. *Ann. Tour. Res.* **2016**, *58*, 159–162. [CrossRef]
20. Shukla, P.; Babin, B.J. Effects of consumer psychographics and store characteristics in influencing shopping value and store switching. *J. Consum. Behav.* **2013**, *12*, 194–203. [CrossRef]
21. Wedel, M.; Kamakura, W.A. *Market Segmentation: Conceptual and Methodological Foundation*; Springer Science & Business Media: Berlin, Germany, 2012.
22. Vyncke, P. Lifestyle segmentation: From attitudes, interests and opinions, to values, aesthetic styles, life visions and media preferences. *Eur. J. Commun.* **2002**, *17*, 445–463. [CrossRef]
23. Gunter, B.; Furnham, A. *Consumer Profiles (Rle Consumer Behaviour): An Introduction to Psychographics*; Routledge: London, UK, 2014.
24. Walters, G.D. *Lifestyle Theory: Past, Present, and Future*; Nova Science Publishers: Commack, NY, USA, 2006.
25. Mitchell, V.W. How to identify psychographic segments: Part 1. *Marketing Intell. Plann.* **1994**, *12*, 4–10. [CrossRef]
26. Bruwer, J.; Li, E. Wine-related lifestyle (WRL) market segmentation: Demographic and behavioural factors. *J. Wine Res.* **2007**, *18*, 19–34. [CrossRef]
27. Schwartz, S.H. An overview of the Schwartz theory of basic values. *Online Read. Psychol. Cult.* **2012**, *2*, 11. [CrossRef]
28. Lin, C.F. Segmenting customer brand preference: Demographic or psychographic. *J. Product. Brand Manag.* **2002**, *11*, 249–268. [CrossRef]
29. Rokeach, M. *The Nature of Human Values*; Free Press: Detroit, MI, USA, 1973.
30. Kahle, L.R.; Beatty, S.E.; Homer, P. Alternative measurement approaches to consumer values: The list of values (LOV) and values and life style (VALS). *J. Consum. Res.* **1986**, *13*, 405–409. [CrossRef]
31. Sagiv, L.; Schwartz, S.H. Cultural values in organisations: Insights for Europe. *Eur. J. Int. Manag.* **2007**, *1*, 176–190. [CrossRef]
32. Yang, J.; Liu, C.; Teng, M.; Liao, M.; Xiong, H. Buyer targeting optimization: A unified customer segmentation perspective, Big Data. In Proceedings of the 2016 IEEE International Conference on IEEE, Washington, DC, USA, 5–8 December 2016; pp. 1262–1271.

33. Hirsh, J.B.; Kang, S.K.; Bodenhausen, G.V. Personalized persuasion: Tailoring persuasive appeals to recipients' personality traits. *Psychol. Sci.* **2012**, *23*, 578–581. [CrossRef]
34. Fernández-Tobías, I.; Braunhofer, M.; Elahi, M.; Ricci, F.; Cantador, I. Alleviating the new user problem in collaborative filtering by exploiting personality information. *User Model. User Adapt. Interact.* **2016**, *26*, 221–255. [CrossRef]
35. Karumur, R.P.; Nguyen, T.T.; Konstan, J.A. Exploring the value of personality in predicting rating behaviors: A study of category preferences on movielens. In Proceedings of the 10th ACM Conference on Recommender Systems, Boston, MA, USA, 15–19 September 2016; pp. 139–142.
36. Lee, H.J.; Lim, H.; Jolly, L.D.; Lee, J. Consumer lifestyles and adoption of high-technology products: A case of South Korea. *J. Int. Consum. Marketing* **2009**, *21*, 153–167. [CrossRef]
37. Pan, Y.; Luo, L.; Liu, D.; Xu, X.; Shen, W.; Gao, J. How to recommend by online lifestyle tagging (olt). *Int. J. Inf. Technol. Decis. Making* **2014**, *13*, 1183–1209. [CrossRef]
38. Piazza, A.; Zagel, C.; Haeske, J.; Bodendorf, F. Do you like according to your lifestyle? a quantitative analysis of the relation between individual facebook likes and the users' lifestyle. In *Proceedings of the International Conference on Applied Human Factors and Ergonomics*; Springer: Cham, Switzerland, 2017; pp. 128–139.
39. Goldsmith, R.E.; Freiden, J.B.; Kilsheimer, J.C. Social values and female fashion leadership: A cross-cultural study. *Psychol. Marketing* **1993**, *10*, 399–412. [CrossRef]
40. Kim, H.S. Consumer profiles of apparel product involvement and values. *J. Fashion Marketing Manag. Int. J.* **2005**, *9*, 207–220. [CrossRef]
41. Heine, K.; Trommsdorff, V. Practicable value-cascade positioning of luxury fashion brands. In Proceedings of the 9th International Marketing Trends Conference, Venice, Italy, 20–23 January 2010; pp. 21–23.
42. Teck, W.J.; Cyril de Run, E. Consumers' personal values and sales promotion preferences effect on behavioural intention and purchase satisfaction for consumer product. *Asia Pac. J. Marketing Logist.* **2013**, *25*, 70–101. [CrossRef]
43. Fraj, E.; Martinez, E. Environmental values and lifestyles as determining factors of ecological consumer behaviour: An empirical analysis. *J. Consum. Marketing* **2006**, *23*, 133–144. [CrossRef]
44. Padgett, D.; Mulvey, M.S. Differentiation via technology: Strategic positioning of services following the introduction of disruptive technology. *J. Retail.* **2007**, *83*, 375–391. [CrossRef]
45. Wiedmann, K.P.; Hennigs, N.; Siebels, A. Value-based segmentation of luxury consumption behavior. *Psychol. Marketing* **2009**, *26*, 625–651. [CrossRef]
46. Antipov, E.; Pokryshevskaya, E. Applying CHAID for logistic regression diagnostics and classification accuracy improvement. *J. Target. Meas. Anal. Marketing* **2010**, *18*, 109–117. [CrossRef]
47. Reutterer, T.; Mild, A.; Natter, M.; Taudes, A. A dynamic segmentation approach for targeting and customizing direct marketing campaigns. *J. Interact. Marketing* **2006**, *20*, 43–57. [CrossRef]
48. Ge, Y.; Xiong, H.; Zhou, W.; Li, S.; Sahoo, R. Multifocal learning for customer problem analysis. *ACM Trans. Intell. Syst. Technol.* **2011**, *2*, 24. [CrossRef]
49. Schwartz, H.A.; Eichstaedt, J.C.; Kern, M.L.; Dziruizynski, L.; Ramones, S.M.; Agrawal, M.; Shah, A.; Kosinski, M.; Stillwell, D.; Seligman, M.E.P.; et al. Personality, gender, and age in the language of social media: The open-vocabulary approach. *PLoS ONE* **2013**, *8*, e73791. [CrossRef] [PubMed]
50. Gosling, S.D.; Mason, W. Internet research in psychology. *Ann. Rev. Psychol.* **2015**, *66*, 877–902. [CrossRef]
51. Adamopoulos, P.; Ghose, A.; Todri, V. The impact of user personality traits on word of mouth: Text-mining social media platforms. *Inf. Syst. Res.* **2018**, *29*, 612–640. [CrossRef]
52. Bleidorn, W.; Hopwood, C.J. Using machine learning to advance personality assessment and Theory. *Personal. Soc. Psychol. Rev.* **2018**, *23*, 190–203. [CrossRef] [PubMed]
53. Park, G.; Schwartz, H.A.; Eichstaedt, J.C.; Kern, C.J.; Kosinski, M.L.; Stillwell, M.; Ungar, J.D.; Seligman, H.L.; Martin, E.P. Automatic personality assessment through social media language. *J. Personal. Soc. Psychol.* **2015**, *108*, 934. [CrossRef] [PubMed]
54. Grunert, K.G.; Perrea, T.; Zhou, Y.; Huang, G.; Sorensen, B.T.; Krystallis, A. Is food-related lifestyle (FRL) able to reveal food consumption patterns in non-Western cultural environments? Its adaptation and application in urban China. *Appetite* **2011**, *56*, 357–367. [CrossRef]
55. Casidy, M.R.; Tsarenko, Y. Predicting brand preferences: An examination of the predictive power of consumer personality and values in the Australian fashion market. *J. Fashion Marketing Manag. Int. J.* **2009**, *13*, 358–371. [CrossRef]
56. Mulyanegara, R.C. The relationship between market orientation, brand orientation and perceived benefits in the non-profit sector: A customer-perceived paradigm. *J. Strateg. Marketing* **2011**, *19*, 429–441. [CrossRef]

57. Yankelovich, D.; Meer, D. Rediscovering market segmentation. *Harvard Bus. Rev.* **2006**, *84*, 122.
58. Sinha, P.K.; Uniyal, D.P. Using observational research for behavioural segmentation of shoppers. *J. Retail. Consum. Serv.* **2005**, *12*, 35–48. [CrossRef]
59. Oly Ndubisi, N.; Tung, M.C. Awareness and usage of promotional tools by Malaysian consumers: The case of low involvement products. *Manag. Res. News* **2006**, *29*, 28–40. [CrossRef]
60. Wang, W.; Li, Y.; Huang, Y.; Liu, H.; Zhang, T. A method for identifying the mood states of social network users based on cyber psychometrics. *Future Internet* **2017**, *9*, 22. [CrossRef]
61. Pero, S.; Huettner, A. Affect analysis of text using fuzzy semantic typing. In *IEEE Transactionsons on Fuzzy Systems*; IEEE: New York, NY, USA, 2001; pp. 483–496.
62. Turian, J.; Ratinov, L.; Bengio, Y. Word representations: A simple and general method for semi-supervised learning. In Proceedings of the 48th Annual Meeting of the Association for Computational Linguistics, Uppsala, Sweden, 11–16 July 2010; pp. 384–394.
63. "Gensim_ models.word2vec – Word2vec embeddings". (2019, May 13). Available online: https://radimrehurek.com/gensim/models/word2vec.html (accessed on 18 February 2019).
64. McAuley, J.; Targett, C.; Shi, Q.; van den Hengel, A. Image-based recommendations on styles and substitutes. In Proceedings of the 38th International ACM SIGIR Conference on Research and Development in Information Retrieval, Santiago, Chile, 9–13 August 2015; pp. 43–52.
65. Müllensiefen, D.; Hennig, C.; Howells, H. Using clustering of rankings to explain brand preferences with personality and socio-demographic variables. *J. Appl. Statistics* **2018**, *45*, 1009–1029. [CrossRef]
66. Levin, N.; Zahavi, J. Predictive modeling using segmentation. *J. Interact. Marketing* **2001**, *15*, 2–22. [CrossRef]
67. Cui, D.; Curry, D. Prediction in marketing using the support vector machine. *Marketing Sci.* **2005**, *24*, 595–615. [CrossRef]
68. Cortez, P.; Cerdeira, A.; Almeida, F.; Matos, T.; Reis, J. Modeling wine preferences by data mining from physicochemical properties. *Decis. Support Syst.* **2009**, *47*, 547–553. [CrossRef]
69. Sharang, A.; Rao, C. Using machine learning for medium frequency derivative portfolio trading. *arXiv* **2015**, arXiv:1512.06228.
70. Esteva, A.; Robicquet, A.; Ramsundar, B.; Kuleshov, V.; DePristo, M.; Chou, K.; Cui, C.; Corrado, G.; Thurn, S.; Dean, J. A guide to deep learning in healthcare. *Nat. Med.* **2019**, *25*, 24. [CrossRef] [PubMed]
71. Vellido, A.; Lisboa, P.J.G.; Meehan, K. Segmentation of the on-line shopping market using neural networks. *Exp. Syst. Appl.* **1999**, *17*, 303–314. [CrossRef]
72. Boone, D.S.; Roehm, M. Retail segmentation using artificial neural networks. *Int. J. Res. Marketing* **2002**, *19*, 287–301. [CrossRef]
73. Arnoux, P.H.; Xu, A.; Boyette, N.; Mahmud, J.; Akkiraju, R.; Sinha, V. 25 Tweets to Know You: A New Model to Predict Personality with Social Media. In Proceedings of the Eleventh International AAAI Conference on Web and Social Media, Montréal, QC, Canada, 15–18 May 2017.
74. Perkins, J. *Python 3 Text Processing with NLTK 3 Cookbook*; Packt Publishing Ltd.: Birmingham, UK, 2014.
75. Guyon, I.; Weston, J.; Barnhill, S.; Vapnik, V. Gene selection for cancer classification using support vector machines. *Mach. Learn.* **2002**, *46*, 389–422. [CrossRef]
76. Bedo, J.; Sanderson, C.; Kowalczyk, A. An efficient alternative to svm based recursive feature elimination with applications in natural language processing and bioinformatics. In *Australasian Joint Conference on Artificial Intelligence*; Springer: Berlin, Germany, 2006; pp. 170–180.
77. Shmueli, G. To explain or to predict? *Statistics Sci.* **2010**, *25*, 289–310. [CrossRef]
78. Zheng, L.; Noroozi, V.; Yu, P.S. Joint deep modeling of users and items using reviews for recommendation. In Proceedings of the Tenth ACM International Conference on Web Search and Data Mining, ACM, Cambridge, UK, 6–10 February 2017; pp. 425–434.

© 2019 by the authors. Licensee MDPI, Basel, Switzerland. This article is an open access article distributed under the terms and conditions of the Creative Commons Attribution (CC BY) license (http://creativecommons.org/licenses/by/4.0/).

Article

Classification of Cyber-Aggression Cases Applying Machine Learning

Guadalupe Obdulia Gutiérrez-Esparza [1,2], Maite Vallejo-Allende [2] and José Hernández-Torruco [3,*]

1. Cátedras CONACYT Consejo Nacional de Ciencia y Tecnología, Ciudad de México 08400, Mexico; ggutierreze@conacyt.mx
2. Instituto Nacional de Cardiología Ignacio Chávez, Ciudad de México 14080, Mexico; maite_vallejo@yahoo.com.mx
3. División Académica de Informática y Sistemas, Universidad Juárez Autónoma de Tabasco, Cunduacán, Tabasco 86690, Mexico
* Correspondence: jose.hernandezt@ujat.mx; Tel.: +52-993-2077-904

Received: 26 March 2019; Accepted: 26 April 2019; Published: 2 May 2019

Abstract: The adoption of electronic social networks as an essential way of communication has become one of the most dangerous methods to hurt people's feelings. The Internet and the proliferation of this kind of virtual community have caused severe negative consequences to the welfare of society, creating a social problem identified as cyber-aggression, or in some cases called cyber-bullying. This paper presents research to classify situations of cyber-aggression on social networks, specifically for Spanish-language users of Mexico. We applied Random Forest, Variable Importance Measures (VIMs), and OneR to support the classification of offensive comments in three particular cases of cyber-aggression: racism, violence based on sexual orientation, and violence against women. Experimental results with OneR improve the comment classification process of the three cyber-aggression cases, with more than 90% accuracy. The accurate classification of cyber-aggression comments can help to take measures to diminish this phenomenon.

Keywords: cyber-aggression; sentiment analysis; random forest; racism; violence based on sexual orientation; violence against women; social networks

---

1. Introduction

The growing access to combined telecommunication along with the increase of electronic social network adoption has granted users a convenient method of sharing posts and comments on the Internet. However, even if this is an improvement in human communication, this environment also has provided proper conditions resulting in serious negative consequences to the welfare of society, due to a type of user who posts offensive comments and does not care about the psychological impact of his/her words, harming other users feelings. This phenomenon is called cyber-aggression [1]. Cyber-aggression is a frequently used keyword in the literature to describe a wide range of offensive behaviors other than cyber-bullying [2–5].

Unfortunately, this problem has spread into a wide variety of mass media; Bauman [6] says that some of the most-used digital media for cyber-aggression are: social networks (e.g., Facebook and Twitter), short message services, forums, trash-polling sites, blogs, video sharing websites, and chat rooms, among others. Their accessibility and fast adoption are a double-edged sword because it is impossible to have moderators to keep an eye on every post made and filter its content. Therefore, cyber-aggression [7] has become a threat to society's welfare, generating electronic violence.

When cyber-aggression is constant, then it becomes cyber-bullying, mainly characterized by the invasion of privacy, harassment, and use of obscene language against one user [8], in most of the cases against a minor [9]. Unlike bullying, cyber-bullying can happen 24/7, and the consequences for the victim can be more dramatic, because this not only creates insecurity, trust issues, and depression, but can also create suicidal thoughts [10] with fatal consequences. Both cyber-bullying and cyber-aggression can harm a user by mocking and ridiculing them an indefinite number of times [11,12]; therefore it is crucial to support the increase of detection of these social problems. According to Ditch The Label and their Cyber-Bullying Survey [13] some of the social networks with the highest number of users that have reported cases of cyber-aggression are Facebook, Twitter, Instagram, YouTube, and Ask.fm.

Currently, there is a wide variety of research aiming to mitigate cyber-bullying. Those from psychology or social sciences aim to detect criminal behavior and provide prevention strategies [14], and those from the field of computer science seek to develop effective techniques or tools to support the prevention of cyber-aggression and cyber-bullying [15]. Nevertheless, it is still complicated to identify a unique behavior pattern of aggression. It is also necessary to highlight that most of the research is carried out by English-speaking countries, so it is difficult to obtain free resources on the Internet for the development of tools that allow the analysis of comments written in Spanish. The Spanish language has an extensive vocabulary with multiple expressions, words that vary in different Spanish-speaking countries, synonyms and colloquialisms that are different in different countries and regions. For this reason, we consider it prudent to focus on the Spanish language of Mexico. However, we do not rule out the possibility of extending this research in other Spanish-speaking countries.

There are different types of cyber-aggression. Bauman [6] identifies and describes some forms of cyber-bullying that are used to attack victims by digital means, such as flaming, harassment, denigration, masquerading, outing, trickery, social exclusion, and cyber-stalking. Peter [16] highlights death threats, homophobia, sexual acts, the threat of physical violence, and damage to existing relationships, among others. Ringrose and Walker [17,18] indicate that women and children are the most vulnerable groups in case of cyber-aggression.

In Mexico, there has been a wave of hate crimes; in 2014, the World Health Organization reported that Mexico occupies second place in the world of hate crimes for factors such as "fear". In Mexico, the case of violence based on sexual orientation is one of the most vulnerable groups, according to the National Survey on Discrimination [19]. As found in an investigation by a group of civil society organizations [20], in Mexico and Latin America, 84% of lesbian, gay, bisexual, and transgender (LGBT) population have been verbally harassed. On the other hand, violence against women is an exponential problem that in Mexico has affected young women between 18 and 30 [21]. Mexico has a public organization known as Instituto Nacional de Estadística y Geografía (INEGI) [22,23], responsible for regulating and coordinating the National System of Statistical and Geographic Information, as well as conducting a national census. INEGI [22] has reported that nine million Mexicans have suffered at least one incident of digital violence in one of its different forms. In 2016, INEGI reported that of the approximately 61.5 million women in Mexico, at least 63% aged 15 and older had experienced acts of violence. Racism is another problem in Mexico that has been growing through offensive messages related to "the wall" with the United States (US), the mobilization of immigrant caravans and discrimination by skin color. According to an INEGI study, skin color can affect a person's job growth possibilities, as well as their socioeconomic status [23].

The American Psychological Association [24] says that different combinations of contextual factors such as gender, race, social class, and other sources of identity can result in different coping styles, and urges psychologists to learn guidelines for psychological practice with people with different problems (e.g., lesbian, gay, and bisexual patients).

For many years, problems from various research areas have been addressed through Artificial Intelligence (AI) techniques. AI is a discipline that emerged in the 1950s and is basically defined as the construction of automated models that can solve real problems by emulating human intelligence. Some recent applications include the following: in [25], two chemical biodegradability prediction models are evaluated against another commonly used biodegradability model; Li et al. [26] applied Multiscale Sample Entropy (MSE), Multiscale Permutation Entropy (MPE), and Multiscale Fuzzy

Entropy (MFE) feature extraction methods along with Support Vector Machines (SVMs) classifier to analyze Motor Imagery EEG (MI-EEG) data; Li and coworkers [27] proposed the new Temperature Sensor Clustering Method for thermal error modeling of machine tools, then the weight coefficient in the distance matrix and the number of the clusters (groups) were optimized by a genetic algorithm (GA); in [28], fuzzy theory and a genetic algorithm are combined to design a Motor Diagnosis System for rotor failures; the aim in [29] is to design a new method to predict click-through rate in Internet advertising based on a Deep Neural Network; Ocaña and coworkers [30] proposed the evolutionary algorithm TS-MBFOA (Two-Swim Modified Bacterial Foraging Optimization Algorithm) and proved a real problem that seeks to optimize the synthesis of a four-bar mechanism in a mechatronic systems.

In this research, we used IA algorithms to classify comments in three cases of cyber-aggression: racism, violence based on sexual orientation, and violence against women. The accurate identification of cybernetic aggression cases is the first step of a process to reduce the incidence of this phenomenon. We applied IA techniques, specifically Random Forest, Variable Importance Measures (VIMs), and OneR. The comments used to create the data set were collected from Facebook, considering specific news related to the cyber-aggression cases included in this study.

In line with the aims of this study, we propose to answer the following research questions:

1. Is it possible to create a model of automatic detection of cyber-aggression cases with high precision? To answer this question, we will experiment with two classifiers of different approaches and compare their performance.
2. What are the terms that allow detection of cyber-aggression cases included in this work effectively? We will seek the answer to this question using methods to identify the relevant features for each cyber-aggression case.

The present work is organized as follows: Section 2 presents a review of related research works, Section 3 describes the materials and methods used in this research, Section 4 shows the architecture of the proposed computational model, Section 5 describes the experiments and results obtained, and the last section concludes the article.

2. Related Research

Due to the numerous victims of cyber-bullying and the fatal consequences it causes, today there is a need to study this phenomenon in terms of its detection, prevention, and mitigation. The consequences of cyber-bullying are worrisome when victims cannot cope with the emotional stress of abusive, threatening, humiliating, and aggressive messages. Presently there are several types of research to avoid or reduce online violence; even so, it is necessary to develop more precise techniques or online tools to support the victims.

Raisi [31] proposes a model to detect offensive comments on social networks, in order to intervene by filtering or advising those involved. To train this model, they used comments with offensive words from Twitter and Ask.fm. Other authors [8,32] have developed conversation systems, based on intelligent agents that allow supportive emotional feedback to victims who suffer from cyber-bullying. Reynolds [33] proposed a system to detect cyber-bullying in the Formspring social network, based on the recognition of violence patterns in the user posts, through the analysis of offensive words; in addition, it uses a ranking level of the detected threat. Likewise, it obtained an accuracy of 81.7% with J48 decision trees.

Ptaszynski [34] describes in his research the development of an online application in Japan, for school staff and parents with the duty of detecting inappropriate content on unofficial secondary websites. The goal is to report cases of cyber-bullying to federal authorities; in this work they used SVMs and got 79.9% accuracy. Rybnicek [12] proposes an application for Facebook to protect minor users from cyber-bullying and sex-teasing. This application aims to analyze the content of images and videos, as well as the activity of the user to record changes in behavior. In another study, a collection of bad words was made using 3915 published messages tracked from the website Formspring.me. The accuracy obtained in this study was only 58.5% [35].

Different studies fight against cyber-bullying by supporting the classification of situations, topics or types of it. For example, at the Massachusetts Institute of Technology, [35] developed a system to detect cyber-bullying in YouTube video comments. The system can identify the topic of the message, such as sexuality, race, and intelligence. The overall success of this experiment was 66.7% accuracy using SVMs. Similarly, the study carried out by Nandhini [36] proposes a system to detect cyber-bullying activities and classify them as flaming, harassment, racism, and terrorism. The author uses a fuzzy classification rule; however, the accuracy of the results are very low (around 40%), but he increased the efficiency of classifier up to 90% using a series of rules. In the same way, Chen [37] proposes an architecture to detect offensive content and identify potential offensive users in social media. The system achieves an accuracy of 98.24% in sentence offensive detection and an accuracy of 77.9% in user offensiveness detection. Similar work to this is presented by Sood [38], in which comments were tagged from a news site using Amazon's Mechanical Turk to create a profanity-labeled data set. In this study, they use SVMs with a profanity list-based and Levenshtein edit distance tool, getting an accuracy of 90%.

In addition to studies that aim to detect messages with offensive content or classify messages into types of cyber-bullying, other studies are trying to prove that a system dedicated to taking care of user behavior can reduce situations of cyber-bullying. Bosse [8] performed an experiment consisting of a normative multi-agent game with children 6 to 12 years old. In the experiment, the author highlights a particular case: a girl who, regardless of the agent's warnings, continued to violate rules and was removed from the game. However, she changed her attitude and began to follow the rules of the game. Through this research, the author shows that in the long term, the system manages to reduce the number of rule violations. Therefore it is possible to affirm that research with technological proposals using sentiment analysis, text mining, multi-agent, or other AI techniques can support the reduction of violence online.

3. Materials, Methods and Metrics

3.1. Materials

Study data. Since we did not find a Spanish data set for the study, we had to collect data through the Facebook API from relevant news in Latin America related to three cases of cyber-aggression: racism, violence based on sexual orientation, and violence against women. We collected 5000 comments; however, we used only 2000, those free of spam (spam characterizes in this study as texts with rare characters, images of expression or humorous thought such as memes, empty spaces, or comments unrelated to the problem). We then grouped the comments (instances) as follows: 700 comments about violence based on sexual orientation, 700 comments about violence against women and 600 racist comments.

Labeling process. In the literature, we found that some researchers [33,38–40] used the web service of Amazon's Mechanical Turk and paid for anonymous online workers to manually label comments, reviews, words, images, among others. However, in [41], tagging by workers from Amazon's Mechanical Turk showed that at least 2.5% of reviews, classified as "no cyber-bullying" should be tagged as "cyber-bullying", for this reason, they sought support from graduate students and some undergraduates of psychology.

Due to the above, we decided to use a group of three teachers with experience in machine-learning algorithms supported by psychologists with experience in evaluation and intervention in cases of bullying in high schools to manually tag the comments. The psychologists explained to professors when a comment is considered offensive in the three cases of cyber-aggression and how to label the comment according to the predefined range of values. So, the purpose of the labeling process was to add an offensive value to each comment considering the case of cyber-aggression and a predefined numerical scale. In the cases of comments with violence against women we used a scale from zero to two, with the lowest value of those comments the least offensive. For comments about violence based on sexual orientation, a scale of values of four to six was used, in the same way with the lowest value for the least offensive comments. Finally, a scale of eight to ten was used for racist comments, with the lowest value for the least offensive comments and the highest value for the most. As a result, we obtained the data set of offensive comments and then used them in the feature-selection procedure and training process. This data set consisted of two columns; the first column contained the instance or comment and second the offensive value according to each comment. We describe, in the following section, the algorithms and methods used in this research for the feature-selection procedure and training process.

3.2. Methods

Random Forest. The Random Forest classification algorithm, developed by Breiman [42], is a set of decision trees that outputs a predictive value and is robust against over-fitting. This algorithm has two parameters, mtry and ntree, which may vary to improve its performance. The first represents the number of input variables chosen at random in each division and the second the number of trees.

Random Forest obtains a class vote of each tree and proceeds to classify according to the vote of the majority. We show the functioning of Random Forest in Algorithm 1.

Algorithm 1: Random Forest for classification
---
1. Let N be the number of training cases, and M the number of variables.
2. Select the value of m, in order to determine the decision in a given node, which must be less than the value of M.
3. Vary and choose N for the tree (completing all training cases) and use the rest of the test cases to estimate the error.
4. For each node of the tree, randomly choose m. Obtain the vote of each tree and classify according to the vote of the majority.

In classification processes, the default value for m is calculated by $\sqrt{p}$ or $[\log 2\ p]$, where p is the number of features. Random Forest provides the VIMs method, through which it is possible to rank the importance of variables in regression or classification problems.

Variable Importance Measures (VIMs). VIMs, based on CART classification trees [43], allows calculation of the weight to identify key attributes or features. In VIMs there are two embedded methods for measuring variable importance proposed by Breiman: Mean Decrease Impurity importance (MDI) and Mean Decrease Accuracy (MDA). MDI ranks each feature importance as the sum over the number of splits that include the feature, proportionally to the number of samples it splits; this method is used in classification cases and applied the Gini index to measure the node impurity [44,45]. Random Forest also uses the Gini index for determining a final class in each tree. MDA is also called the permutation importance and is based on out-of-bag (OOB) [46] samples to measure accuracy; this method is suitable for regression problems [43]. In this work we used MDI with Gini index to classify comments in cases of cyber-aggression, which is defined by:

$$G(t) = 1 - \sum_{k=1}^{Q} p^2(k|t) \qquad (1)$$

where Q is the number of classes, $p(k|t)$ is the estimated class probability for feature t or node t in a decision tree and k is an output class.

At the end of the process, VIMs calculate a score for each variable, normalized by the standard deviation. The higher the score, the more important the feature. In this work, we used R [47] and applied the *randomForest* [48] library to develop the classification model.

OneR. OneR stands for One Rule. OneR is a rule-based classification algorithm. A classification rule has the form: if attribute1 <relational operator> value1 <logical operator> attribute2 <relational operator> value2 <...> then decision-value. OneR generates one rule for each predictor in the data set, and among them all, it selects the rule with the smallest total error as the only one rule [49]. One advantage of rules is that they are simple for humans to interpret.

3.3. Evaluation Metrics

Confusion Matrix. It shows the performance of the prediction model, comparing the results of the predictive model against real values.

Accuracy (ACC). This is the evaluation of the predictive model performance. In binary classification, the accuracy is calculated by dividing the number of correctly identified cases among the total cases. It is computed by:

$$ACC = \frac{TP + TN}{T + N} \qquad (2)$$

Sensitivity (TRP). This metric measures the proportion of true positives which were correctly classified and is calculated by:

$$TRP = \frac{TP}{TP + FN} \qquad (3)$$

Specificity (SPC). This metric measures the proportion of true negatives, which were correctly classified. It is given by:

$$SPC = \frac{TN}{FP + TN} \qquad (4)$$

where TP = True Positive, TN = True Negative, FP = False Positive and FN = False Negative, in Equations (2)–(4).

Kappa statistic. This method introduced by Cohen [50] and Ben-David [51] is used to measure the accuracy of machine-learning algorithms. Kappa measures the agreement between the classifier itself and the ground truth corrected by the effect of the agreement between them by chance. This method is defined by:

$$K = \frac{P_o + P_c}{1 - P_c} \qquad (5)$$

where P_o is the proportion of agreement between the classifier and the ground truth, P_c is the proportion of agreement expected between the classifier and the ground truth by chance.

4. Experimental Procedure

4.1. Random Forest

This section describes the steps followed to carry out the experimental procedure with Random Forest (see Figure 1).

Phase 1. Comment Extraction. We detail this step in the Materials section (study data).

Phase 2. Feature Selection. In this phase, we applied the VIMs method to identify the key attributes for each cyber-aggression case [43]. The feature-selection process is shown in Figure 2. We applied this process for each cyber-aggression case. The first step, after we cleaned the comments, was to separate them according to a numerical range, as follows: 0–2 for comments related to violence based on sexual orientation, 4–6 for those about violence against women, and 8–10 for those related to racism. Then, we identified the frequent terms to eliminate those repeated and not important. Later, we partitioned the set of comments (corpus) using 2/3 for training and 1/3 for testing. At the same time, we applied the VIMs method along with the Gini index to obtain the terms with the higher weight to create the feature corpus. After this process, we configured a vector of 30 seeds, and then we applied the *random.forest.importance* instruction. We executed this process 30 times, considering a different seed to train with different partitions of the training corpus, in order to evaluate the results of Random Forest. We considered high-performance values greater than 70%, and then it was possible to use the corpus of features in the parameter optimization process, which we applied to find the most appropriate value of *mtry*. In this process, we used the *Fselector* library in the R system for statistical computing [47].

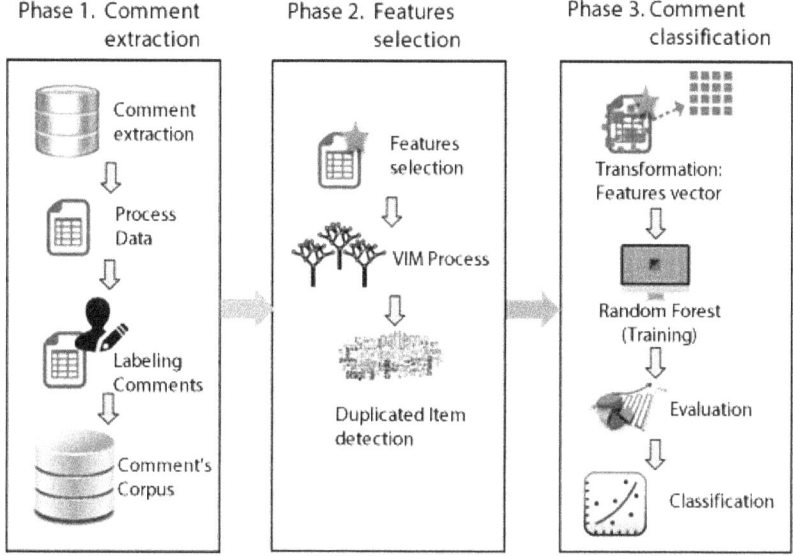

Figure 1. Architecture of the proposed computational model.

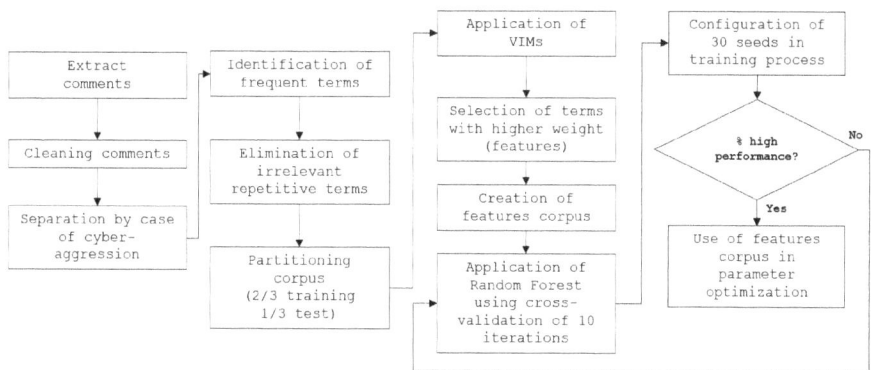

Figure 2. Feature-selection process.

Phase 3. Classification. This step is the training process considering the feature corpus, identified by VIMs (applied in phase 2 of the Model). We used 10-fold cross-validation and Random Forest algorithm to adjust the *mtry* and *ntree* parameters. After the training process, we carried out the classification process using Random Forest, which we executed another 30 times, implementing in each execution a different seed, generated by the Mersenne-Twister method [52]. To test the result of the classification in different executions with different sections of training-sets, a variety of seeds is important. Figure 3 represents the steps followed in this phase.

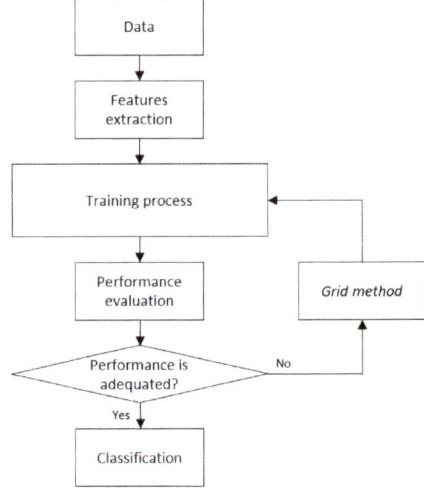

Figure 3. Classification process.

4.2. OneR

This section describes the experimental procedure for OneR classifier.

For each aggression case, we first calculated the frequency of appearance of each term and the average frequency of appearance of all the terms. Then, we selected the terms above this average to use the most frequent terms as predictors. Afterwards, we made a subsetting of the original data set containing only the selected terms and created a new data set. From this new data set, we created three data sets to perform the classification experiments. In the first data set, we kept the LGBT class and marked the other two classes as ALL; in the second we kept the Machismo class and marked the other two as ALL; and also with the third. Finally, we performed 30 independent runs, where we calculated the average values of the metrics used across all the study.

5. Results

5.1. Feature Selection Using VIMs

We used the VIMs method to identify the key features of each cyber-aggression case, as shown in Figure 2 and described in Phase 2 (feature selection). The result of this process (see Table 1) was made in two executions—initial and final. In the initial execution, we showed the results without a cleaning process; in the final execution, the results are shown considering a cleaning process and the application of VIMs method. Column 1, "Case", of Table 1, represents the cyber-aggression case, where VS is violence based on sexual orientation, VM is violence against women and R is racism. The second column, "execution", indicates the initial and final executions from each cyber-aggression case. The third column, "Potential features", presents the arithmetic mean of potential features in each cyber-aggression case. Column 4 shows the arithmetic mean of weights obtained from all terms, where we can see that values from the final execution increase in comparison with the result from the initial execution. In column five, "Maximum weight" represents the maximum weights obtained in each cyber-aggression case and each type of execution; this term appears in Table 2, in the column of Importance from each type of cyber-aggression. The final column, "Minimum weight", is the minimum weights obtained in each case of cyber-aggression and each case of execution.

Table 1. Results of the feature-selection procedure.

Case	Execution	Potential Features	Average Weights	Maximum Weight	Minimum Weight
VS	Initial	33.33%	5.43	29.44	1.00
VS	Final	89.33%	6.44	33.45	0.13
VM	Initial	29.16%	4.33	21.09	0.02
VM	Final	88.49%	5.24	27.04	0.075
R	Initial	48.06%	4.65	23.66	1.00
R	Final	89.10%	6.87	28.04	0.20

The processing and cleaning of comments influenced the results of the final execution, because in the first execution some irrelevant words such as "creo" ("I think" in English) obtained a high value of 11.70, "comentarios" ("comments" in English) obtained a value of 5.36 and "cualquiera" ("anyone" in English) obtained a value of 2.34.

We measured the importance of the terms with VIMs using the Gini index. Table 2 shows the most important terms that were identified by this method, which are those with the highest value. In the case of violence based on sexual orientation, the most important term was "asco" ("disgust") with a value of 33.4539, in the case of violence against women, the most important term was "estúpida" ("stupid") with a value of 27.0473, and in racist comments, the most important term was "basura" ("garbage") with a value of 28.0465. We show in Table 2 an extract of important term results (feature selection) for each cyber-aggression case after the cleaning of comments; some terms include an "*", which is not part of the original term, but was used to censor the offense.

Table 2. Results of the importance of the terms in the feature-selection procedure.

VS Term	Importance	VM Term	Importance	R Term	Importance
asc*	33.4539	est*pida	27.0473	b*sura	28.0465
circo	28.0842	c*chetadas	19.5111	escl*vos	26.3688
dios	25.7459	c*ger	17.7479	color	22.8754
d*generados	24.1965	c*lera	16.3509	ch*nos	20.9468
ab*rración	23.9967	cualquiera	14.4007	as*sinos	19.5273
an*rmal	18.0855	estorban	13.0006	r*za	18.9392
asqu*roso	17.8648	p*rras	12.7073	n*gro	12.4821
en*ermedad	17.6001	cocina	12.6314	en*no	12.2413
depr*vados	14.4772	drama	12.3626	an*mal	10.6022
an*rmales	13.5536	br*ta	11.6915	b*gotonas	10.0998
mar*cas	12.6927	c*lebra	11.0332	cr*stianos	9.2143
est*pidez	11.0387	p*ta	10.3944	pr*etos	7.3209
biblia	10.9030	ab*sivas	10.3768	t*rroristas	6.6306
cabr*nes	10.8184	b*rra	9.5017	c*zar	6.2582
enf*rmos	10.6553	débil	8.9978	m*tar	6.0199
cerrada	10.2412	descuidada	8.7899	enemigos	5.5840
an*males	9.5388	arg*enderas	8.4999	ch*cha	4.6080
arrepientanse	9.1431	asqu*rosas	8.1761	ind*o	4.4520
d*sviados	8.9325	c*gan	7.8184	d*lincuente	4.3617
deberían	8.4167	m*ntenidas	7.6039	gr*ngo	4.3598

Once we identified the features for each problem, we carried out the training process (phase 3), where we evaluated the results using the metrics described in Section 3.3 (evaluation metrics). Finally, we applied the classifier model, if the accuracy exceeded 70%.

5.2. Training Process and Parameter Tuning

To allow the algorithm to be prepared to perform a better classification, we carried out a training process with Random Forest. We selected the 50 features with the highest weight of each problem; also, we performed 10-fold cross-validation and a variation of the value of the mtry parameter using random numbers between zero and 50. We used these random numbers in the same way for each problem. We measured the training performance of the algorithm using the accuracy and Kappa metrics. In Table 3, we show the overall results obtained in this process, where R is racism, VS is violence based on sexual orientation and VM is violence against women. In column two (Mtry), we show the optimal values of the mtry parameter obtained for each case of cyber-aggression, column three presents the accuracy obtained with respect of the value of the mtry parameter, and column four shows the value of Kappa with respect of the value of the mtry parameter. The values used in the mtry parameter were in a range of 23 to 44. Figure 4 shows the complete results for the adjustment of the mtry parameter.

Table 3. Results of the adjustment of the mtry parameter.

Case	Mtry	Accuracy	Kappa
VS	50	0.8211	0.5789
VM	23	0.8327	0.6217
R	44	0.9169	0.7737

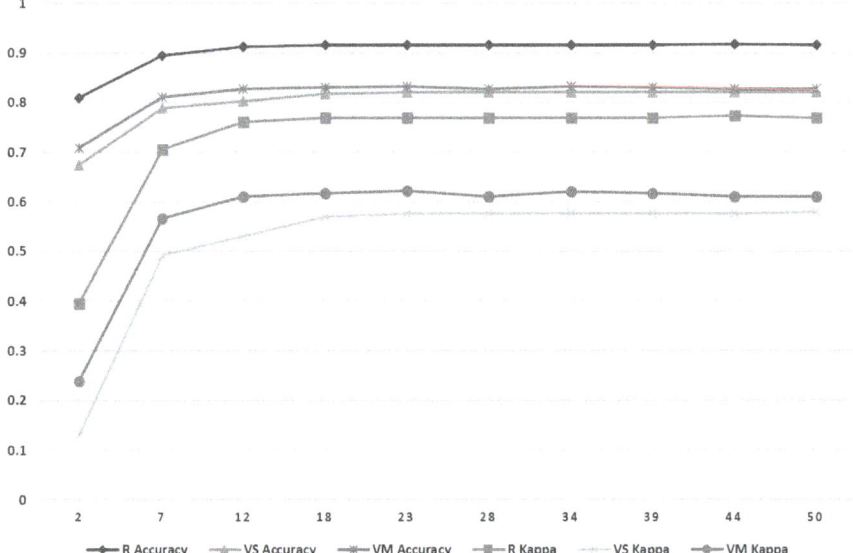

Figure 4. Graph of adjustment of the results of the mtry parameter.

In the case of the ntree parameter, we considered two values: 100 and 200, based on the investigations reported in [53,54]. Oshiro [53] concludes that the variation of the mtry with values that indicate a higher number of trees only increases the computational cost and does not have a significant improvement ratio.

Once we obtained the value of mtry for each case of cyber-aggression, as well as the value of ntree, we carried out the classification process with Random Forest. It is essential to highlight that we made this process with 30 executions varying the training and test set, in order to test the performance of Random Forest with different sections of the data set (explained in phase three of the architecture of the model).

Based on the results obtained with accuracy and confusion matrices, in the 30 executions, metrics such as balanced accuracy, sensitivity, and specificity were obtained. These results are presented in Table 4, as well as the standard deviation (sd) of each metric depending on cyber-aggression case.

As can be seen, the model obtains a better performance in the presented metrics, even though in the three problems it highlights the sensitivity with low values and the specificity with high values. The performance of the model indicates that negative comments were correctly classified; this proves that the results were very close to the expected performance because 95% of the comments in this data set were offensive. Since this work focuses on detecting potential features to classify comments on three problems of cyber-aggression (racism, sexual violence, and violence against women), it was justified in implementing a data set composed primarily of negative comments.

Table 4. Results of comment classification. Standard deviation (sd) is also shown.

Case	Accuracy	Balanced Accuracy	Sensitivity	Specificity
R	0.9030	0.9237	0.6783	0.9883
sd	0.0129	0.0129	0.0466	0.0060
VS	0.8036	0.8313	0.5320	0.9591
sd	0.0162	0.0203	0.0355	0.0115
VM	0.8204	0.8233	0.6325	0.9261
sd	0.0148	0.0187	0.0311	0.0165

5.3. Results with OneR

As Table 5 shows, we found an average frequency of terms for VS aggression of 13.41, 21.61 for VM, and 18.60 for R. This means that for example in VS, terms appear 13.41 times as an average, where the most frequent term is "mar*c*nes", with 29 appearances. In VM, terms appear 21.61 times as an average, the most frequent term being "vieja" with 76 appearances. Finally, in R, terms appear 18.60 times as an average, with "negros" as the most frequent with 63 appearances. We show the selected terms in the same table, i.e., those with a frequency of appearance greater than the average frequency of appearance of all the terms. Figure 5, known as word clouds, show the importance of the terms for each type of aggression. The bigger the term in the figure, the more relevant it is. These pictures match the terms shown in Table 5. As expected, the most common offensive terms in Spanish are present. The most common offensive words are "maric*nes", "asco", "put*s", "gays", "gay", "mi*rda" for VS; "vieja", "pinch*", "mujeres", "put*", "viejas", "pinch*s" for VM, and "negros", "mi*rda", "indio", "negro", "judíos", "musulmanes" for R. We used the terms from Table 5 as input to OneR classifier to build the classification models for each cyber-aggression case.

Table 5. Selected terms by their average frequency for each type of aggression.

Case	Average Frequency of Terms	Selected Terms (Quantitiy)
VS	13.41	maric*nes, asco, put*s, gays, gay, mi*rda, hombres, pinch*s, jot*s, maric*s, dios, mundo, gente, homosexuales, ser, respeto, put* (17)
VM	21.61	vieja, pinch*, mujeres, put*, viejas, pinch*s, bien, mujer, p*nd*ja, put*s, solo, así (12)
R	18.6	negros, mi*rda, indio, negro, judíos, musulmanes, malditos, p*nd*ejo, mexicanos, pinch* (10)

Figure 5. Word clouds for each cyber-aggression case. (a) Violence based on sexual orientation (VS). (b) Violence against women (VM). (c) Racism (R).

Table 6 shows the classification results with OneR using only the most relevant terms found, as described in the Experimental Procedure section. We show the results averaged over 30 independent runs. We also show the standard deviation (sd). We found an accuracy above 0.90 for all types of aggression. Moreover, all metrics reached 0.90 in most cases. Results prove the efficiency of the methodology applied.

Table 6. Classification results using only the most relevant terms for each type of aggression. The results are shown averaged over 30 runs. Standard deviation (sd) is also shown.

Case	Accuracy	Balanced Accuracy	Sensitivity	Specificity	Kappa
R	0.9623	0.9569	0.9712	0.9426	0.9119
sd	0.0080	0.0103	0.0071	0.0184	0.0188
VS	0.9525	0.9192	0.8383	1.0000	0.8795
sd	0.0102	0.0174	0.0348	0.0000	0.0274
VM	0.9545	0.9615	0.9274	0.9957	0.9066
sd	0.0099	0.0112	0.0138	0.0237	0.0206

In Table 7, we show the best classification rule generated by OneR for each type of aggression. Also, the number of correctly classified instances and the balanced accuracy in the test set represents the effectiveness of each rule. We can see that OneR selected the same most relevant terms for each type of aggression. Some irrelevant terms in Spanish were included, such as "así", "bien", "solo", among others. They represent adverbs and prepositions instead of nouns or adjectives.

Table 7. Best rule for each aggression case.

Case	VS		VM		R	
Classification Rule	termino:		termino:		termino:	
	asco	-> lgbt	asco	-> ALL	asco	-> ALL
	dios	-> lgbt	dios	-> ALL	dios	-> ALL
	g*y	-> lgbt	g*y	-> ALL	g*y	-> ALL
	g*ys	-> lgbt	g*ys	-> ALL	g*ys	-> ALL
	gente	-> lgbt	gente	-> ALL	gente	-> ALL
	hombres	-> lgbt	hombres	-> ALL	hombres	-> ALL
	homosexuales	-> lgbt	homosexuales	-> ALL	homosexuales	-> ALL
	j*tos	-> lgbt	j*tos	-> ALL	j*tos	-> ALL
	m*ric*s	-> lgbt	m*ric*s	-> ALL	m*ric*s	-> ALL
	m*ric*nes	-> lgbt	m*ric*nes	-> ALL	m*ric*nes	-> ALL
	mi*rda	-> ALL	mi*rda	-> ALL	mi*rda	-> racismo
	mundo	-> lgbt	mundo	-> ALL	mundo	-> ALL
	pinch*s	-> ALL	pinch*s	-> machismo	pinch*s	-> ALL
	p*ta	-> ALL	p*ta	-> machismo	p*ta	-> ALL
	p*tos	-> lgbt	p*tos	-> ALL	p*tos	-> ALL
	respeto	-> lgbt	respeto	-> ALL	respeto	-> ALL
	ser	-> lgbt	ser	-> ALL	ser	-> ALL
	así	-> ALL	así	-> machismo	así	-> ALL
	bien	-> ALL	bien	-> machismo	bien	-> ALL
	mujer	-> ALL	mujer	-> machismo	mujer	-> ALL
	mujeres	-> ALL	mujeres	-> machismo	mujeres	-> ALL
	p*nd*ja	-> ALL	p*nd*ja	-> machismo	p*nd*ja	-> ALL
	p*nche	-> ALL	p*nche	-> machismo	p*nche	-> ALL
	p*tas	-> ALL	p*tas	-> machismo	p*tas	-> ALL
	solo	-> ALL	solo	-> machismo	solo	-> ALL
	vieja	-> ALL	vieja	-> machismo	vieja	-> ALL
	viejas	-> ALL	viejas	-> machismo	viejas	-> ALL
	indio	-> ALL	indio	-> ALL	indio	-> racismo
	judíos	-> ALL	judíos	-> ALL	judíos	-> racismo
	malditos	-> ALL	malditos	-> ALL	malditos	-> racismo
	mexicanos	-> ALL	mexicanos	-> ALL	mexicanos	-> racismo
	musulmanes	-> ALL	musulmanes	-> ALL	musulmanes	-> racismo
	negro	-> ALL	negro	-> ALL	negro	-> racismo
	negros	-> ALL	negros	-> ALL	negros	-> racismo
	p*nd*jo	-> ALL	p*nd*jo	-> ALL	p*nd*jo	-> racismo
Correct instances	772/818		775/818		782/818	
Balanced Accuracy	0.9458		0.9776		0.9774	

6. Discussion and Conclusions

Cyber-aggression has increased negatively as the use of social networks increases, which is why in this work we have sought to develop computational tools analyzing offensive comments and classified into three categories.

There are already a variety of software tools or applications that operate under AI techniques to detect offensive comments, filter them or send messages of support to the victim. However, it is still necessary to improve the performance of these tools to get more effective predictions. The development of tools that work in the Spanish language is also required, since most of the research targets English-speaking countries, which is why it is difficult to obtain resources for algorithm training, such as data sets, lexicons, corpora, among others. Moreover, it is crucial to consider certain idioms or colloquialisms of the region where the model applies, so the translation of the available resources of the Web is not always convenient. Therefore it is necessary to create resources in the Spanish language. According to the need for a data set of offensive comments in Spanish, and considering the example of other related research, the authors have created their own data sets of comments using social networks such as Twitter [55], Formspring.me [35], Facebook [56], Ask.fm [31] and others where cyber-bullying has been increasing [57].

We decided to create a data set of offensive comments using Facebook. At first, Twitter was used to extract comments on these three cases of cyber-aggression, but most of the comments were irrelevant. The Twitter API allows download of comments using a hashtag, but there are few users who make offensive comments and use a hashtag to identify the comment. For this reason, we decided to use news on Facebook about marriage between people of the same sex, publications about women's triumphs, or reports of physical abuse and news about Donald Trump's wall. Besides, Facebook is the most-used social network in Mexico [58]. We believe that a gathering of more relevant news comments such as abortion, adoption by same-sex couples, feminicides and other related news, not only from Mexico but also from Latin America, as well as the inclusion of experts who study the Spanish language and colloquialisms, can improve the performance of the classifier.

This paper describes the development of a model to classify cyber-aggression cases applying Random Forest and OneR. We seek to initially impact Mexico, where there has been a wave of hate crimes. Our contribution is as follows: (1) We created a data set with cyber-aggression cases from social networks. (2) We focused on cyber-aggression cases in our native language, i.e., Spanish. (3) Specifically, we were interested in the most representative types of cyber-aggression in our country of origin, Mexico. (4) We identified the most relevant terms for the detection of cyber-aggression cases included in the study. (5) We created an automatic detection model of cyber-aggression cases with high precision and interpretable by the human being (rule-based).

The results obtained in this work with Random Forest support the identification of relevant features to classify offensive comments into three cases: racism, violence based on sexual orientation, and violence against women. Nevertheless, OneR outperformed Random Forest in identifying types of cyber-aggression, in addition to providing a simple classification rule with the most relevant terms for each type of aggression. Even when we obtained high-performance classification models with this particular data, it is essential to highlight that the classifiers used in this study have a better performance classifying offensive comments against the LGBT population and racists comments. Therefore the exploration of other machine-learning techniques and the continuous update of the offensive data set may not be ruled out in future work, thus allowing the analysis of another kind of cyber-aggression case, e.g., those suffered by children. Also, it is important to continue improving the feature-selection process. On the other hand, building an automatic labeling system for offensive comments made by social networks users, and thus minimizing human error, will be of great help. Finally, we will seek to identify the victims of cyber-aggression to provide them with psychological attention according to the case of harassment that they suffer.

Author Contributions: Conceptualization—G.O.G.-E.; Methodology—G.O.G.-E., J.H.-T.; Software—G.O.G.-E.; Supervision—M.V.-A.; Validation—G.O.G.-E.; formal analysis—J.H.-T.; writing—original draft preparation—G.O.G.-E., M.V.-A.; writing—review and editing—G.O.G.-E., M.V.-A.

Funding: This research received no external funding.

Acknowledgments: We want to extend our appreciation to Consejo Nacional de Ciencia y Tecnología (CONACYT) (National Council for Science and Technology) under the 'Catedras CONACYT' programme, No. 1591.

Conflicts of Interest: The authors declare no conflicts of interest.

References

1. Kowalski, R.M.; Limber, S.P.; Limber, S.; Agatston, P.W. *Cyberbullying: Bullying in the Digital Age*; John Wiley & Sons: Hoboken, NJ, USA, 2012.
2. Grigg, D.W. Cyber-aggression: Definition and concept of cyberbullying. *J. Psychol. Couns. Sch.* **2010**, *20*, 143–156. [CrossRef]
3. Kopecký, K.; Szotkowski, R. Cyberbullying, cyber aggression and their impact on the victim—The teacher. *Telemat. Inform.* **2017**, *34*, 506–517. [CrossRef]
4. Corcoran, L.; Guckin, C.; Prentice, G. Cyberbullying or cyber aggression?: A review of existing definitions of cyber-based peer-to-peer aggression. *Societies* **2015**, *5*, 245–255. [CrossRef]
5. Watkins, L.E.; Maldonado, R.C.; DiLillo, D. The Cyber Aggression in Relationships Scale: A new multidimensional measure of technology-based intimate partner aggression. *Assessment* **2018**, *25*, 608–626. [CrossRef] [PubMed]
6. Bauman, S. *Cyberbullying: What Counselors Need to Know*; John Wiley & Sons: Hoboken, NJ, USA, 2014.
7. Fredstrom, B.K.; Adams, R.E.; Gilman, R. Electronic and school-based victimization: Unique contexts for adjustment difficulties during adolescence. *J. Youth Adolesc.* **2011**, *40*, 405–415. [CrossRef]
8. Bosse, T.; Stam, S. A normative agent system to prevent cyberbullying. In Proceedings of the 2011 IEEE/WIC/ACM International Conferences on Web Intelligence and Intelligent Agent Technology-Volume 02. IEEE Computer Society, Lyon, France, 22–27 August 2011; pp. 425–430.
9. Kowalski, R.M.; Giumetti, G.W.; Schroeder, A.N.; Lattanner, M.R. Bullying in the digital age: A critical review and meta-analysis of cyberbullying research among youth. *Psychol. Bull.* **2014**, *140*, 1073. [CrossRef] [PubMed]
10. Williams, M.L.; Pearson, O. Hate Crime and Bullying in the Age of Social Media. 2016. Available online: http://orca.cf.ac.uk/88865/1/Cyber-Hate-and-Bullying-Post-Conference-Report_English_pdf.pdf (accessed on 25 May 2018).
11. Del Rey, R.; Casas, J.A.; Ortega, R. The ConRed Program, an evidence-based practice. *Comunicar* **2012**, *20*, 129–138. [CrossRef]
12. Rybnicek, M.; Poisel, R.; Tjoa, S. Facebook watchdog: a research agenda for detecting online grooming and bullying activities. In Proceedings of the 2013 IEEE International Conference on Systems, Man, and Cybernetics (SMC), Manchester, UK, 13–16 October 2013; pp. 2854–2859.
13. DitchTheLabel.org. The Annual Cyberbullying Survey. 2013. Available online: https://www.ditchthelabel.org/wp-content/uploads/2016/07/cyberbullying2013.pdf (accessed on 25 May 2018).
14. Turan, N.; Polat, O.; Karapirli, M.; Uysal, C.; Turan, S.G. The new violence type of the era: Cyber bullying among university students: Violence among university students. *Neurol. Psychiatry Brain Res.* **2011**, *17*, 21–26. [CrossRef]
15. Van Royen, K.; Poels, K.; Daelemans, W.; Vandebosch, H. Automatic monitoring of cyberbullying on social networking sites: From technological feasibility to desirability. *Telemat. Inform.* **2015**, *32*, 89–97. [CrossRef]
16. Smith, P.K. Cyberbullying and cyber aggression. In *Handbook of School Violence and School Safety*; Routledge: London, UK, 2012; pp. 111–121.
17. Ringrose, J.; Gill, R.; Livingstone, S.; Harvey, L. *A Qualitative Study of Children, Young People and 'Sexting': A Report Prepared for the NSPCC*; National Society for the Prevention of Cruelty to Children: London, UK, 2012.
18. Walker, S.; Sanci, L.; Temple-Smith, M. Sexting: Young women's and men's views on its nature and origins. *J. Adolesc. Health* **2013**, *52*, 697–701. [CrossRef]
19. CONAPRED. National Survey on Discrimination. 2010. Available online: https://www.conapred.org.mx/userfiles/files/ENADIS-2010-Eng-OverallResults-NoAccss.pdf (accessed on 1 March 2019).
20. FUNDACIONARCOIRIS. 2a Encuesta Nacional sobre Violencia Escolar basada en la Orientación Sexual, Identidad y Expresión de Género hacia Estudiantes LGBT en México. 2018. Available online: www.fundacionarcoiris.org.mx (accessed on 19 February 2019).
21. INMUJERES. Ciberacoso. 2018. Available online: https://www.gob.mx/inmujeres/articulos/ciberacoso?idiom=es (accessed on 19 February 2019).

22. INEGI. Microdatos del Modulo sobre Ciberacoso (MOCIBA). 2015. http://www.beta.inegi.org.mx/contenidos/proyectos/investigacion/ciberacoso/2015/doc/mociba2015_principales_resultados.pdf (accessed on 16 November 2018).
23. INEGI. Módulo de Movilidad Social Intergeneracional (MMSI). 2016. https://www.inegi.org.mx/programas/mmsi/2016/ (accessed on 11 February 2019).
24. American Psychological Association. Guidelines for psychological practice with lesbian, gay, and bisexual clients. *Am. Psychol.* **2012**, *67*, 10. [CrossRef]
25. Baker, J.R.; Gamberger, D.; Mihelcic, J.R.; Sabljic, A. Evaluation of Artificial Intelligence Based Models for Chemical Biodegradability Prediction. *Molecules* **2004**, *9*, 989–1003. [CrossRef] [PubMed]
26. Li, M.a.; Liu, H.n.; Zhu, W.; Yang, J.F. Applying Improved Multiscale Fuzzy Entropy for Feature Extraction of MI-EEG. *Appl. Sci.* **2017**, *7*, 92. [CrossRef]
27. Li, F.; Li, T.; Wang, H.; Jiang, Y. A Temperature Sensor Clustering Method for Thermal Error Modeling of Heavy Milling Machine Tools. *Appl. Sci.* **2017**, *7*, 82. [CrossRef]
28. Kuo, C.C.; Liu, C.H.; Chang, H.C.; Lin, K.J. Implementation of a Motor Diagnosis System for Rotor Failure Using Genetic Algorithm and Fuzzy Classification. *Appl. Sci.* **2017**, *7*, 31. [CrossRef]
29. Wang, Q.; Liu, F.; Xing, S.; Zhao, X. A New Approach for Advertising CTR Prediction Based on Deep Neural Network via Attention Mechanism. *Comput. Math. Methods Med.* **2018**, *2018*, 1–11. [CrossRef]
30. Hernández-Ocaña, B.; Pozos-Parra, M.D.P.; Mezura-Montes, E.; Portilla-Flores, E.A.; Vega-Alvarado, E.; Calva-Yáñez, M.B. Two-Swim Operators in the Modified Bacterial Foraging Algorithm for the Optimal Synthesis of Four-Bar Mechanisms. *Comput. Intell. Neurosci.* **2016**, *2016*, 1–18. [CrossRef]
31. Raisi, E.; Huang, B. Cyberbullying identification using participant-vocabulary consistency. *arXiv* **2016**, arXiv:1606.08084.
32. Van der Zwaan, J.M.; Dignum, V.; Jonker, C.M. A conversation model enabling intelligent agents to give emotional support. In *Modern Advances in Intelligent Systems and Tools*; Springer: Berlin/Heidelberg, Germany, 2012; pp. 47–52.
33. Reynolds, K.; Kontostathis, A.; Edwards, L. Using machine learning to detect cyberbullying. In Proceedings of the 2011 10th International Conference on Machine learning and applications and workshops (ICMLA), Honolulu, HI, USA, 18–21 December 2011; Volume 2, pp. 241–244.
34. Ptaszynski, M.; Dybala, P.; Matsuba, T.; Masui, F.; Rzepka, R.; Araki, K. Machine learning and affect analysis against cyber-bullying. In Proceedings of the 36th AISB, Leicester, UK, 29 March–1 April 2010; pp. 7–16.
35. Dinakar, K.; Reichart, R.; Lieberman, H. Modeling the detection of Textual Cyberbullying. *Soc. Mob. Web* **2011**, *11*, 11–17.
36. Nandhini, B.S.; Sheeba, J. Online social network bullying detection using intelligence techniques. *Procedia Comput. Sci.* **2015**, *45*, 485–492. [CrossRef]
37. Chen, Y.; Zhou, Y.; Zhu, S.; Xu, H. Detecting offensive language in social media to protect adolescent online safety. In Proceedings of the 2012 International Conference on Privacy, Security, Risk and Trust and 2012 International Confernece on Social Computing, Amsterdam, The Netherlands, 3–5 September 2012; pp. 71–80.
38. Sood, S.O.; Antin, J.; Churchill, E.F. Using Crowdsourcing to Improve Profanity Detection. In Proceedings of the AAAI Spring Symposium: Wisdom of the Crowd, Palo Alto, CA, USA, 26–28 March 2012; Volume 12, p. 6.
39. Rosa, H.; Pereira, N.; Ribeiro, R.; Ferreira, P.; Carvalho, J.; Oliveira, S.; Coheur, L.; Paulino, P.; Simão, A.V.; Trancoso, I. Automatic cyberbullying detection: A systematic review. *Comput. Hum. Behav.* **2019**, *93*, 333–345. [CrossRef]
40. Potha, N.; Maragoudakis, M. Cyberbullying detection using time series modeling. In Proceedings of the 2014 IEEE International Conference on Data Mining Workshop, Shenzhen, China, 14 December 2014; pp. 373–382.
41. Ptaszyński, M.; Leliwa, G.; Piech, M.; Smywiński-Pohl, A. Cyberbullying Detection–Technical Report 2/2018, Department of Computer Science AGH, University of Science and Technology. *arXiv* **2018**, arXiv:1808.00926.
42. Breiman, L. Random forests. *Mach. Learn.* **2001**, *45*, 5–32. [CrossRef]
43. Hjerpe, A. *Computing Random Forests Variable Importance Measures (VIM) on Mixed Numerical and Categorical Data*; KTH, School of Computer Science and Communication (CSC): Stockholm, Sweden, 2016.
44. Han, H.; Guo, X.; Yu, H. Variable selection using mean decrease accuracy and mean decrease gini based on random forest. In Proceedings of the 2016 7th IEEE International Conference on Software Engineering and Service Science (ICSESS), Beijing, China, 26–28 August 2016; pp. 219–224.

45. Breiman, L.; Cutler, A. Breiman and Cutler's Random Forests for Classification and Regression. 2018. Available online: https://cran.r-project.org/web/packages/randomForest/randomForest.pdf (accessed on 12 April 2019).
46. Breiman, L. *Out-Of-Bag Estimation*; University of California: Berkeley CA, USA, 1996. Available online: https://www.stat.berkeley.edu/~breiman/OOBestimation.pdf (accessed on 12 April 2019).
47. R Core Team. *R: A Language and Environment for Statistical Computing*; R Foundation for Statistical Computing: Vienna, Austria, 2013.
48. Liaw, A.; Wiener, M. Classification and regression by randomForest. *R News* **2002**, *2*, 18–22.
49. Witten, I.; Frank, E.; Hall, M. *Practical Machine Learning Tools and Techniques*; Morgan Kaufmann: Burlington, MA, USA, 2011.
50. Cohen, J. A coefficient of agreement for nominal scales. *Educ. Psychol. Meas.* **1960**, *20*, 37–46. [CrossRef]
51. Ben-David, A. About the relationship between ROC curves and Cohen's kappa. *Eng. Appl. Artif. Intell.* **2008**, *21*, 874–882. [CrossRef]
52. Matsumoto, M.; Nishimura, T. Mersenne twister: A 623-dimensionally equidistributed uniform pseudo-random number generator. *ACM Trans. Model. Comput. Simul.* **1998**, *8*, 3–30. [CrossRef]
53. Oshiro, T.M.; Perez, P.S.; Baranauskas, J.A. How many trees in a random forest? In *International Workshop on Machine Learning and Data Mining in Pattern Recognition*; Springer: Berlin/Heidelberg, Germany, 2012; pp. 154–168.
54. Torruco, J.H. Descriptive and Predictive Models of Guillain-Barré Syndrome Based on Clinical Data Using Machine Learning Algorithms. Ph.D. Thesis, Universidad Juárez Autónoma de Tabasco, Tabasco, México, 2015.
55. Sanchez, H.; Kumar, S. Twitter bullying detection. *Ser. NSDI* **2011**, *12*, 15.
56. Hamouda, A.E.D.A.; El-taher, F. Sentiment analyzer for arabic comments system. *Int. J. Adv. Comput. Sci. Appl.* **2013**, *4*. [CrossRef]
57. Van Hee, C.; Jacobs, G.; Emmery, C.; Desmet, B.; Lefever, E.; Verhoeven, B.; De Pauw, G.; Daelemans, W.; Hoste, V. Automatic detection of cyberbullying in social media text. *PLoS ONE* **2018**, *13*, e0203794. [CrossRef] [PubMed]
58. INFOTEC. 13o. Estudio sobre los Hábitos de los Usuarios de Internet en México. 2018. Available online: https://www.infotec.mx/work/models/infotec/Resource/1012/6/images/Estudio_Habitos_Usuarios_2017.pdf (accessed on 11 February 2019).

© 2019 by the authors. Licensee MDPI, Basel, Switzerland. This article is an open access article distributed under the terms and conditions of the Creative Commons Attribution (CC BY) license (http://creativecommons.org/licenses/by/4.0/).

MDPI
St. Alban-Anlage 66
4052 Basel
Switzerland
Tel. +41 61 683 77 34
Fax +41 61 302 89 18
www.mdpi.com

Applied Sciences Editorial Office
E-mail: applsci@mdpi.com
www.mdpi.com/journal/applsci

www.ingramcontent.com/pod-product-compliance
Lightning Source LLC
LaVergne TN
LVHW071955080526
838202LV00064B/6754